How to Build and Modify HIGH-PERFORMANCE Manual Transmissions

Paul Cangialosi

S-A DESIGN

CarTech®

Edited by: Paul Johnson
Designed by: Monica Bahr

ISBN-13 978-1-932494-05-1

Printed in China

CarTech®

39966 Grand Avenue
North Branch, MN 55056
Telephone (651) 277-1200 • (800) 551-4754 • Fax: (651) 277-1203
www.cartechbooks.com

OVERSEAS DISTRIBUTION BY:

Brooklands Books Ltd.
P.O. Box 146, Cobham, Surrey, KT11 1LG, England
Telephone 01932 865051 • Fax 01932 868803
www.brooklands-books.com

Brooklands Books Aus.
3/37-39 Green Street, Banksmeadow, NSW 2019, Australia
Telephone 2 9695 7055 • Fax 2 9695 7355

DEDICATION

I probably would not be writing this book if not for Fred Lewandowski. Fred allowed me as a young teen to channel my energy in a creative direction. He also gave me the best automotive education that no school could ever match. I never would have thought back then that my knowledge gained from working with this man would give me a lifetime career.

At an early age, my father taught me how to build model cars. He was an extraordinary model builder. Model building teaches patience, craftsmanship, pride in one's work, importance of reading instructions, and creative vision. For that, I thank you, Dad.

I blame my Uncle Joe for his involvement in my addiction to manual transmissions. He allowed me to spend summers at his house reading car magazines and catalogs. I drove his 1965 Falcon Sprint (a 4-speed, of course), up and down his driveway, which subsequently led to my first clutch installation, shifter adjustment, and initiation to the smell of gear lube.

I thank my mother and grandparents for enduring years of transmission parts and disassembled autos scattered across the backyard and sometimes on the kitchen table. How did they ever put up with it?!

My wife Cora has stood by my side and also has had to endure a missing husband for countless days, nights, and sometimes weeks. It is only through her strength and conviction in me that I am able to write this book. Although I have an editor for this book, Cora has become the editor for my life.

I also thank the late Steve Hendrickson and CarTech Books for believing in me. He made a life-long dream come true.

I thank Brian Higgins of SK speed for his countless hours aiding me with parts, statistics, and his life's knowledge.

There are so many new acquaintances and friends that have developed as a result of this project. Thank you for your time, knowledge, and belief in me.

Shift Fast,
Paul Cangialosi

PREFACE

My first introduction into the workings of a transmission and why one is needed came with the acquisition of my first bicycle. It was a 20-inch single-speed bike. I could only pedal to a point, in which no matter how fast I pedaled, acceleration flattened out. I was able to achieve a higher "top end" by changing out the pedal sprocket to a larger diameter. However, this made my initial acceleration suffer. My uncle Joe had this "English Racer" bicycle, which as children we were instructed never to touch. It had five speeds! With the help of some tools and extra hubs, I was able to stick his sprocket assemblies on my 20-inch bike. This new sprocket arrangement gave me the acceleration and top end I needed, and it was probably the fastest 20-inch bike in town. This was my first introduction to the importance of gearing and transmissions.

My journey into this madness started in the late 1970s. An "A-level Journeyman" taught me how to repair engines and automatic transmissions. Four-speeds, back then, were only in hot rods or muscle cars, so as a general repair mechanic I rarely saw them. When one did finally surface for repair, I was hooked. I think the physical aspect sets building and repairing these types of transmissions apart from automatics. The end user has to shift it physically. That person's enjoyment or frustration of his or her shifting experience is directly related to how well you build the transmission. In fact, each time a person pulls a clean, smooth shift, he or she may be actually thinking of how great of a job the rebuilder did.

Over a course of 27 years, I've met many people with the same addiction towards gears, and interestingly, there are not that many. I estimate there are about fewer than 30 people in this U.S. who specialize in manual transmissions alone, and probably no more than 15 who do it on a full-time basis. Most are still hobbyists and work a full-time job for real income. Some stick to one type of transmission and become specialists in that particular design. There are others who just specialize in doing modifications, such as making lightweight 2-speeds out of 4-speeds for circle track racing. It starts by accumulating mass quantities of similar transmissions. It is not uncommon for a transmission specialist to have a few hundred of the same gearbox. It takes years of building and collecting parts to actually decipher what types of castings, ratios, and even hardware were used in the production of certain makes and models. Gear manufacturing in the U.S. doesn't exist on a large scale anymore. Major car manufacturers purchase the bulk of standard shift transmissions from Mexico, Japan, and Germany. Corvettes and Mustangs all use Mexican-made transmissions. There are fewer than 10 specialty shops in this country that actually design and build new high-performance or racing components for this market today.

The original concept of this book was based on a smaller manual I called *Setting Standards*. Released in the late 1980s, the book mainly provided answers to common questions related to transmissions. These questions came from 14-plus car shows that I attended every year. What I gathered was that there is a certain type of person who is drawn to manual-shift transmissions. It could be the smell of 90wt gear lube or the attraction of mechanical components working together. I also noticed that most automatic transmission or engine builders rarely venture into manual transmission territory. The precision setup of a geartrain can be intimidating. So, the goal of this book is to break down the mystique of the gearbox. This book will help you understand design concepts, theory, building techniques, and history of the manual-shift transmission. If you decide to purchase or rebuild a standard transmission the explanations set forth here should aid the reader in making better choices and build confidence.

Today's internal combustion engine requires a transmission to effectively transmit power. Electric motors or steam engines do not. These get power from an outside source. Using steam or electricity to power them, they can generate torque at zero rpm. Also, both steam engines and electric motors can easily operate in reverse direction without the use of a transmission. This cannot be easily accomplished with an internal combustion engine. If you try putting a load on a typical gasoline engine at very low RPM, it will stall. When most people first learn to drive with a manual-shift transmission, they usually stall out the engine until the technique of slipping the clutch is mastered. Getting back to riding a 5-speed bicycle, I always tell people to associate your legs as being the engine. It will be very easy to start off in first gear. Your legs won't have to work as hard. You may be able to start off in fifth gear, but you will have to work very hard to initially start. The extra work of starting in fifth gear eventually strains and fatigues your legs. The same is true for an engine. When a high gear is selected at low RPM, the engine lugs in that gear because it is overworked. Lugging causes unwanted wear and tear on engine components. So what you learn is that even though a transmission is required, the proper use of gear ratios for a specific application is just as important.

Manual transmissions today are far more complex then those built in the 1950s or 1960s. The objective of this book is to touch upon high-performance transmissions. Most high-performance transmissions today are based on the older rear-wheel-drive designs. This simple design incorporates an input shaft and an output shaft that attaches to a driveshaft that spins your rear axle. You may hear of the term "transaxle." Transaxles, mounted in the rear of the car, incorporate a transmission and rear axle gearing as a complete unit. Tremec's T56 6-speed came as a standard rear-wheel-drive transmission in the Camaro, but it was a transaxle in the Corvette. A front-wheel-drive axle does the opposite. The transmission and final drive are mounted in the front, and axles drive the front wheels. Many cars are now equipped with all-wheel drive. All-wheel drives are usually computer controlled to give optimum traction and power to the wheel that needs it the most. All-wheel drive should not be confused with 4-wheel drive. Typical, 4-wheel-drive (4x4) vehicles, such as a Jeep for example, use a conventional rear-wheel-drive transmission with a "transfer case" bolted to the back of it. The transfer case would then send power to the front wheels and also act as a link to the rear wheels.

Your basic manual-shift transmission has an input shaft, countergear, speed gears, and an output shaft. Typically, power comes into the input shaft, goes down through the countergear, then up to a speed gear and out through the output shaft. Selection of a particular gear is accomplished by locking the speed gear to the output shaft. The amount of speed gears in a transmission determines if the transmission is a 3-speed, 4-speed, 5-speed, etc. Each pair of gears in a transmission is called a set. Each set can have a specific gear ratio. One turn of the drive gear in the set can create multiple turns, one turn, or less than one turn of the driven gear. This is the basis of a gear ratio. The difference between one gear set's ratio as opposed to another is called a "spread" or "drop." This is expressed as a percentage.

After dealing with many mechanics and transmission shops over the years, many people still don't understand what a gear ratio is, how it is calculated, and how important proper use of ratios are. Gear ratios can be used in different operational modes. The modes are underdrive, overdrive, direct, reverse, and neutral. In underdrive mode, the input spins more than the output. In overdrive mode, the output spins more than the input. In direct mode, both input and output are either locked together or the overall ratio is 1:1. Neutral allows both input and output to be independent of one another. Reverse causes the output to spin in the opposite direction of the input and is always an underdrive ratio. We determine gear ratios by using simple division and multiplication.

This book focuses on high-performance manual transmissions. The transmissions overviewed in this book are based on current popularity as well as historical demand for parts and service. There are no 3-speed transmissions featured because they are no longer considered a high-performance transmission. Light-duty 4-speeds, such as the GM Saginaw, have been omitted from the build section as well.

The transmissions I decided to include were ones that I'm asked about everyday. Manual transmissions pretty much function in the same way. The cases and size of the gears may change, but the basic principles are very similar. Therefore, I included chapters on theory as well as common rebuilding techniques that apply to most of them. If a transmission is not overviewed in this book, one should be able to use these basic procedures to tackle most transmission jobs.

A brief history of each make and model is given. This is not meant to be a restorer's guide; however, certain popular models will have more historical data as well as ID information.

Shifters and clutch information were added since they are something you will have to address when working on your transmission. To help round out your newfound knowledge of manual transmissions, I included a chapter on high-end race transmissions rarely seen or discussed in most publications.

Although this book should answer most questions one has in regards to building manual transmissions, support for any additional information can be obtained online. Just go to the support forums at www.5speeds.com.

INTRODUCTION

You should congratulate yourself for investing in this book. In this time when people expect free information from the Internet, fewer and fewer people are investing in comprehensive information found in manuals and books that make complete buildups possible. Professional repair shops as well as individuals for some odd reason no longer feel they have to purchase overhaul manuals for cars or individual components they need to work on. Every workday includes a few phone calls to me from people who purchased the wrong transmission, wrong parts or got misinformation from the Internet. A great deal of research went into this publication. That research is based on 25 years in this business as well as interacting with some of the best people in the standard transmission field.

A great deal of transmissions produced in the 1940s to late 1950s shared similar design concepts. More cars are mass-produced today than in the 1940s, and it is interesting to note that there are more surplus gears still sitting on shelves for transmissions made during the 1940s and 1950s. The reason is that they broke more often, and spare parts were incorporated into the design concept.

During the 1960s through the end of the 1970s, the manual-shift transmission business was hot. There was a great deal of horsepower being generated, and transmissions were not holding up to abuse that well. The OEM (Original Equipment Manufacturers) spent the first part of this decade improving on their original designs to accept the increasing horsepower levels. If you were involved with gearbox building during this time, you witnessed first hand how changes in castings or alloys altered strength. The significant designs created at this time are still the foundation for today's transmissions. I've taken a great deal of time to contact friends in the industry and speak to builders who are "in the know" to help document some of these changes.

By 1975 catalytic converters and lower horsepower levels started to prevail, and the OEM no longer had to accept responsibility for broken transmissions. If you purchased a Corvette in 1975 that had an advertised 190 hp and decided to modify the engine to make 375 hp, it was now your problem if you exploded the transmission. The period from 1975 to 1982 is what I call the "dry period." This was a time during which the aftermarket and OEMs did very little. The cost to design and manufacture new gears or castings was extremely high and limited only to a few shops. The economic recession in the U.S. didn't help matters as well. Most of the shops that did work for the OEM in the past also did the bulk of the professional race work for NASCAR or sanctioned drag race organizations.

By 1983 the IBM PC started to change the way people worked. Computer Aided Design (CAD) programs started to find there way into smaller machine shops. This was the start of "New Design Period." Many new designs were created and tweaked out during this period. Newer style alloys, new ways to form gear teeth as part of a forging process, and the use of friction paper in synchro rings instead of traditional bronze were developed. Powdered metal processes and strong computer-designed die-cast cases were some of the advances. This period from 1983 to 1996 represents a time in which software designers were able to give engineers better tools to improve reliability, gas mileage and incorporate fewer components.

We are currently in the "high-tech period." By 1996 3D modeling software started to also become affordable, and this empowered a designer to truly see his vision. You can now look at your design concept from various angles, throw all sorts of "what if" data around, and thus create a more stable design. Once again, cars are now producing impressive power levels and are able to meet emissions requirements. This has resulted in new demand for aftermarket transmission parts. It seems that improving product reliability is no longer a priority for the OEMs and larger manufacturers. Each year when I go to the Performance Racing Industry show, I notice that the bigger manufacturers seem to have the same old designs and have very few people interested in what they have to offer. In contrast, there is usually another company with just two employees actually designing and building much more sophisticated transmissions. Needless to say, their booths are overcrowded.

Every week someone asks me if I have any NOS (new old stock) gears. They have a hard time embracing new technology. For some odd reason they do not believe transmission technology has improved in 30 years. In fact, the quality of steels related to gears and bearings *has* greatly improved because the alloys are more consistent and pure. Gears are now machined more accurately than equivalent parts produced 40 years ago. A small shop now can produce a quality piece that exceeds anything a major manufacturer made in 1960.

The future of the manual transmission business is based on the past. When researching how many new companies have started up in this field over the past 10 years, I am shocked to see that only a few have survived. Some die off quick simply because they don't have the foundation of the older technology. I've witnessed first hand how horrible gearbox designs fail because they are based on software rather than instinctual knowledge. The modern gearbox now has a computer or brain, and the brain is supposed to shift for you. Maybe this is a necessity in Formula 1 racing; however, it seems to me that we are taking the fun out of driving and also are redefining what the term "manual shift" means! So consider this a purist's book. I won't bore you with computer ECM troubleshooting, solenoid resistances, or places to plug your laptop in.

CHAPTER 1

THEORY OF MANUAL TRANSMISSIONS

Whether you are dealing with sprocket teeth on a bike or gear teeth on a transmission gear, the simple formula for calculating a gear ratio is: **Driven / Drive = Ratio.**

So if the drive gear has 10 teeth and the driven gear 20 teeth, the ratio equals 2. This is expressed as a 2:1 or a 2.00 ratio. The drive gear must make two turns to make the driven gear turn once. Any ratio with a number greater than 1 is an underdrive ratio. A ratio less than 1 becomes an overdrive ratio. If the drive gear has 40 teeth and the driven has 30, the ratio becomes .75. This is expressed as a .75:1 ratio. Sometimes overdrive ratios are expressed as a percentage. If the ratio is .75, the difference between .75 and 1.0 is .25 or 25 percent. Therefore, a .75 overdrive ratio is often called a 25% overdrive.

Since your average manual shift transmission contains more than one pair of gears, the same formula holds true for each drive and driven set. The ratio of each drive and driven set is multiplied by each other to give the final ratio. **The formula is driven divided by drive multiplied by driven divided by drive.** When looking at the power-flow pictures for first-gear mode, you will see that the power comes into the input shaft (drive) down to the countergear (driven) then from the countergear (drive) to the first-speed gear (driven). You can now figure your overall gear ratio. Here is an example: Your input shaft has 21 teeth. The mating driven section of the countergear has 25 teeth. The first-gear section of the countergear has 17 teeth, and the first-speed gear has 36 teeth. The result is 25 divided by 21 multiplied by 36 divided by 17. This yields a first-gear ratio of 2.52:1.

Gear ratios are a very important aspect of transmission selection and transmission design. Ratios can help

Power flow is shown in neutral.

Power flow is arranged in first gear.

Power flow is pictured in second gear.

Power flows through in third gear.

Power flow is positioned in fourth or direct gear.

Power flow is pictured in fifth gear.

Power flow is shown in reverse.

These two gears both have the same tooth counts, but one has a smaller diameter. The larger gear has thicker teeth.

determine proper application as well as the torque capacity of a transmission. Two areas often overlooked when selecting a transmission are gear ratio and center-to-center distance. The center-to-center distance is the distance between the centerlines of the upper and lower geartrains. When visualizing why a center-to-center distance is important I'll give an extreme example. You can have two gear sets. Both sets have a 20-tooth driven gear and 10-tooth drive gear; however, one set has a center-to-center of 1 inch while the other a center-to-center of 3-1/2 inches. The set with the larger center-to-center will obviously have larger teeth and bigger gears. This yields a stronger transmission, but with a heavy geartrain. The transmission with the small center-to-center may shift easier, because of the lighter mass of the geartrain, but it will be weaker. The proper selection of a transmission for a particular application is based on this idea, thus a car with a 150-hp engine that weighs 1,800 lbs would be robbed of performance if the engine had to turn a transmission designed for a 600-hp 3,000-lb vehicle.

Torque capacity seems to be the latest buzz word. Published torque ratings of transmissions are often misleading as well as misunderstood. As mentioned above, longer center-to-center distances improve capacity. Gear ratios affect capacity in several ways. A 3:1 ratio compared to a 2:1 ratio along the same center-to-center distance is usually weaker by nature of gear design. Fewer teeth are needed to obtain a larger reduction. A smaller diameter gear with fewer teeth is weaker than a gear larger in diameter with more teeth. There are some trade offs. The more teeth you put on a gear within a fixed diameter, the finer the pitch of that said gear. This finer pitch results in a gear-tooth profile with a thinner cross section as opposed to a gear the same diameter with fewer teeth. You can actually improve the strength of the gear by giving it a bigger diameter with fewer teeth. In mass-produced transmissions, the manufacturer very rarely makes a separate profile for each gear set. That is what the specialty performance shops do.

So what does a published rating of 300 ft-lbs of torque really mean? A torque rating is not a static rating. This means you don't stick a breaker bar and lock the transmission in two gears and apply 300 ft-lbs of torque to the input shaft and watch something snap when you exceed the 300 lbs. Consider a torque rating as a life factor of a transmission based on a particular application. After speaking in depth to various engineers from several manufacturers in regards to this issue, you may want to think of a torque rating as being a dynamic rating based on a theoretical life factor. Let's say we feel a transmission should yield good service for 100,000 miles in a car that has an engine that produces 300 ft-lbs of peak torque. Obviously, the transmission is never seeing that engine's peak torque all the time. Actually, unless you have no rear axle, the torque coming out of the engine is absorbed by both the transmission and rear axle. In fact, a dead rear-axle ratio, such as a 3.08 rear, will cause the transmission to load more than a 4.11 rear. You can exceed the rating of the transmission but you will shorten the transmission's life factor. The rating is actually a safe benchmark. Different manufacturers arrive at published ratings using different methods of calculations, making it really impossible to compare one transmission to another. The Tremec T5 5-speed originally designed by BorgWarner had its published ratings actually change when Tremec took over the line. The T5s have their published ratings exceeded every day. The ones that survive probably have better rear-axle gearing, such as a 4.11. Transaxles designed for the 24 Hours of LeMans race are loaded on a dyno to simulate real life loads against engines producing more than 1,000 hp. They are designed to last for that one race. I prefer real world applications and factual data as opposed to hypothetical data any day. So keep an open mind when asking about or reading published ratings.

Today, more and more people are swapping various engines and transmissions into custom projects. Few take the time to understand these basic principles that I've just explained. Much energy, time, and money are wasted trying to adapt Toyota Supra transmissions in Jaguars or 4-cylinder S-10 Truck 5-speeds into a 1968 Camaro. I am asked these "will it work?" questions every week. In some parts of the world, certain transmissions are more readily available than others. Australia has an over abundance of Toyota 5-speeds, thus people are always trying to stick them into everything in sight. The point is to save time and money by selecting the proper application for a project.

Wide and Close Ratios Explained

What does having a wide- or close-ratio manual shift transmission mean? First of all, it has nothing to do with how close or wide your shifter moves within its pattern. That term is actually called "long or short throw" shifter. I'll discuss here a simple method that will show visually what close and wide ratio is.

Ratio 1 – Ratio 2 = Difference
Difference / Ratio 1 = Percentage Drop
***Figure 1 Formula**

There is no particular formula that determines a close- or wide-ratio gearbox. I say this because what was considered wide ratio in the 1960s is considered close ratio today. Confused yet? Since you now know what a gear ratio is, the difference between two gear ratios as mentioned earlier can be called a ratio spread or drop. That drop is a percentage of the previous ratio. **Figure 1** shows the formula. Let's say we have a close-ratio 4-speed Super T10. The ratios are: 2.64 first, 1.61 second, 1.23 third, and 1.00 fourth. The ratio change between the gears is: 39%, 24%, and 19%.

BorgWarner's extra-low-ratio Super T10 had a 2.88 first, 1.75 second, 1.33 third and 1:00 fourth. Notice if you plug these gear ratios into the above formula, you will end up with the following drop: 39%, 24%, and 25%.

Think of gear ratios as a distance between two points. In a typical muscle car, 4-speed application fourth gear is direct or 1:1 ratio. Again, thinking in terms of distance, consider fourth gear or direct as your final destination. The further you get away from 1:1 (direct) the wider the ratio in one sense. The Muncie M21 and M22 close-ratio 4-speeds were the closest ratio 4-speeds ever put into a production car. Their ratios were 2.20 first, 1.64 second, 1.28 third, 1:00 fourth. If you plug those ratios into the formula in Figure 1, you will end up with 25%, 22%, and 22%. Compare these percentage drops to the previous drops found using the Super T10. Notice that although the 2-3 and 3-4 drops of the Super T10 and Muncie are pretty close percentage wise, but the 1-2 drops vary quite a bit. The reason is that the Super T10s have a lower (higher numerical) first gear. The 2.88 and 2.64 ratios are further away from direct than the Muncie's 2.20 first-gear ratio.

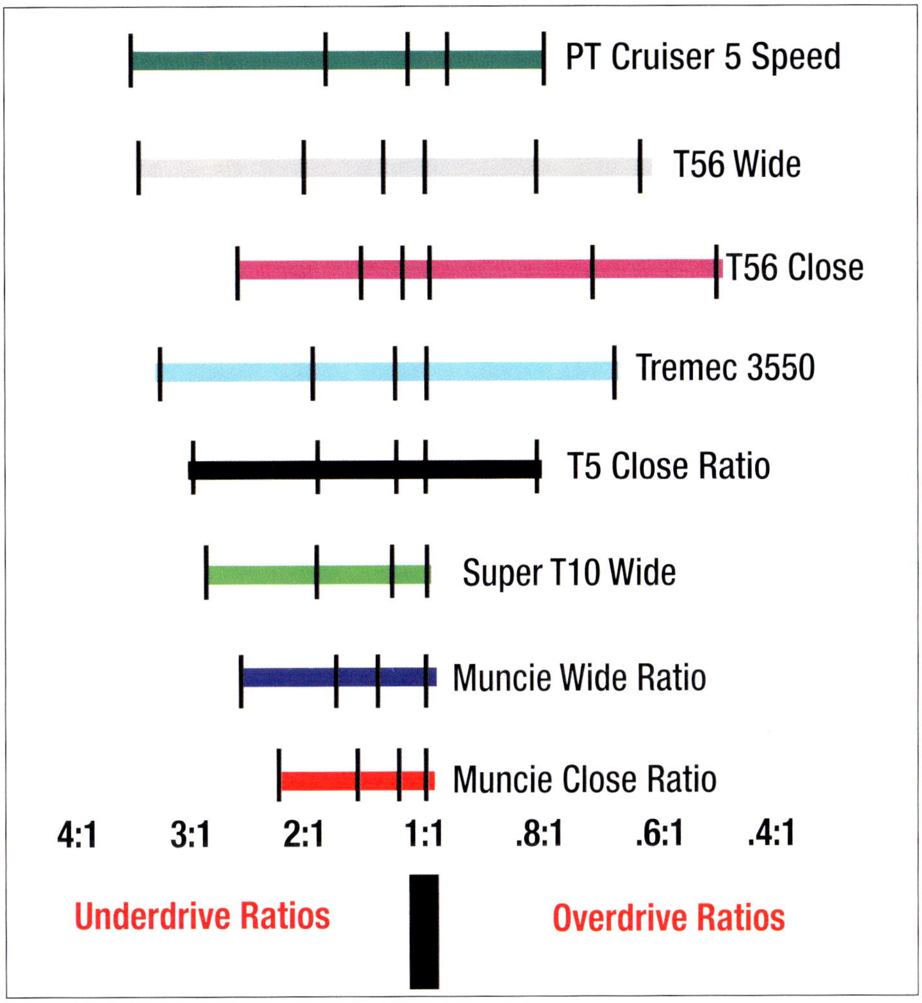

This shows how different ratios compare on paper.

What Makes Close-Ratio Close?

As you have learned, you can't gain distance without losing closeness. The lower the first gear, the wider the 4-speed will be. If one takes a historical look at the close-ratio gearbox, it usually had 25% or lower drops across all gear spreads. The problem is: the Muncie close-ratio box was designed in the 1960s when we really didn't worry much about gas mileage. In order for you to get a "close ratio," a Muncie-equipped car had at least a 3.70 rear-end gear. You needed at least that low of a final drive to get your car moving. The "wide ratio" M20 had a 2.52 first gear. You could get a 3.31 or 3.55 rear with that and gain a little more economy, but it wasn't cool to have what was considered a wide-ratio box in your Corvette. By late 1974, fuel economy and air pollution became considerations, and catalytic converters were mandatory by 1975. Axle ratios had to drop to get the improved gas mileage the Enviromental Protection Agency (EPA) demanded. So in order to get cars moving with 3.08 or 2.88 rear-end gears, cars that had engines with decent amounts of low-end torque were equipped with "close-ratio" transmissions. Larger engines carried transmissions with 2.64 first gears while smaller-displacement engines' first gear ratios were in the 3.0 range. So the newer close-ratio transmissions are wider than the older wide- and close-ratio transmissions!

Doug Nash Corporation had an interesting solution to this dilemma. It developed a 5-speed box with a close-ratio spread like a Muncie M21 with a fifth-gear direct. Again, think of what I said about distance between two points. The more stops you make in your travel to get to your final destination, the closer the distance between your stops. By adding a lower first-speed gear and keeping a direct fifth, you gain more distance, but add an extra stop. The Doug Nash Street 5-speed came with a 3.27 first gear. This allowed drag enthusiasts to use a 3.08 final drive and still get good close ratio acceleration. This worked great.

Newer cars today produce a great deal of low-end torque. Most peak power is made in the 4,500-rpm range. The average 5-speed, such as a T5, is an extremely wide ratio by 1960s standards. The close-ratio T5s used by Ford Motorsport have a 2.95 first, 1.94 second, 1.34 third, 1.00 fourth, and .80 fifth. You do the math with the formula. These new five and six speeds in a sense are really wide-ratio 4-speeds with additional overdrive gears for fifth and sixth speed. Cars can now cruise at 1,800 rpm at 70 mph because the engine's torque curve can handle the load. If you are still thinking in terms of distance, not only are you getting farther away from our final destination of direct drive, but you are now going past your final destination into two levels of overdrive.

Chrysler's new PT Cruiser has a 2.4L 150-hp 4-cylinder engine. The engine produces 162 ft-lbs @ 4,000 rpm, and the 5-speed transaxle has a final drive of 3.94! The first ratio is 3.50, second is 1.96, third is 1.36, fourth is .97, and fifth is .81. I hope you can actually see this is really a 3-speed with two overdrive ratios. The engine really doesn't produce power until it hits 4,000 rpm. Thus, the gearing sort of works, but the car can be boggy at lower RPM.

Goodbye Wide and Close Ratios

The pictured chart shows shift points of some popular transmissions. For the most part, cars are no longer offered with close- or wide-ratio transmission options. If you look at the chart, you will see how by plotting stops of a distance traveled you realize what close or wide ratio may mean. In one sense, if you compare stops, some transmissions have similar distances between one or two stops while others are drastically different. What is close or wide ratio is no longer an issue. Rather, how much percentage drop your engine can handle is the issue.

Road racing events, such as NASCAR- or SCCA-type races, still require transmissions with a close-ratio spread. It is common for some of these cars to have 1.86 first-gear ratios. Plot a gearbox on the above chart with a 1.86 first, 1.59 second, 1.17 third, and 1.00 fourth. How do those results look in comparison to the others? What percentage drops does that same box create using the formula? A low first gear is not needed because road race cars tend to operate at the higher range of RPM and speed with no stopping other than for a pit stop. An ultra close-ratio gearbox has other advantages as well. Since the load changes are not as severe, drivetrain parts tend to live longer. The high-shock loads of wide-ratio transmissions usually cause gearbox failure and rear-axle failure. Less heat is generated in a close-ratio box. A car having a final drive of 2.98 and a direct fourth gear has the same overall ratio as a car with a .80 overdrive fifth and 3.73 final drive. The car with the overdrive uses more HP and generates more heat through the transmission than the direct drive box. However, the 3.73 rear may offer more low-speed punch on turns. My point here is that when it comes to what is right for your car, it is still a matter of trial and error.

Getting your street or race car geared properly produces fantastic results. Jericho, G-Force and Liberty's all make excellent 4-and 5-speed gearboxes for drag-race applications. If you have a track-only car, then sticking a 4.11 or 4.56 rear-end gear and using 4- or 5-speed race gearbox will produce better results than any street 5-speed. When it comes to road racing, find other people using similar engine and rear combinations to yours. Compare how they do against one another, especially if they use a different transmission.

The How and Why of Synchros

Historically, manual transmissions were defined as either progressive or selective. When we shift a transmission, we are disengaging one gear from the mainshaft (also called the output shaft) and engaging another. A progressive transmission has one massive gearset with one shifting mechanism. When you shift you mate this single one-piece set to various gears on the countergear assembly. Thus, it has one lever. A selective transmission has individual gears that are locked to the output. A typical 4-speed transmission has three levers, a lever each for 1-2, 3-4, and for reverse. How we make the transition from one gear to another is accomplished by sliding a gear into another by use of a synchronizer.

Leading edges of gears chip, but overhang once the gears are engaged.

"Sliding gear technology," as I call it, is often misunderstood. Today, very few transmissions use sliding gears, but any old 4-speed still uses a sliding non-synchro reverse gear. The later T5 5-speeds used a sliding reverse and then a synchronized reverse brake. I sell more reverse gear sets because people see chipped teeth and think they need to be replaced. In many cases, they probably don't need to be replaced and will be chipped within a few months anyway. A sliding gear has no way of stopping the gear it is being mated to, so it will usually grind and chip the leading edges. What most people don't realize is that the gear, when fully engaged, overhangs the mating gear. The edge that initially contacts the mating gear slides past it and overhangs on the back side, while the mating gear overhangs on the front side. Consequently, this area is designed to be chipped! Usually, if the gear is chipped no more than one-quarter of the way, the gear will be fine.

Synchronizers make sure the selected gear is locked to the output shaft at the same speed so no grinding occurs. Every performance manual-shift transmission that uses synchronizers follows these rules:

- All gears are in constant mesh; this means that they don't need to slide into one another. They are in fact already spinning as a matched set.
- Although the gears are in constant mesh, they are floating on the output shaft.
- The synchronizer is always locked to the output shaft.
- The synchronizer's job is to lock a particular gear to the output shaft, so power can now flow through the gear and out of the shaft.

Understanding manual transmission theory, components, and operation is the key to successfully diagnosing problems. Many people think they know how synchros work, but usually misdiagnose

Figure A illustrates the upshift mode. The gear is powered, and in order for the slider to mate with the gear it must index ahead of it.

Figure B shows the downshift mode. The slider is powered, and in order for the slider to mate with the gear, it must index behind it. Whether you are upshifting or downshifting, a "blockout" condition can occur in which wear of the synchro assembly allows the ring and slider to meet point to point.

because of a poor understanding of the theory. Let's see if we can crack this mystery once and for all.

Synchronize, as defined by Webster's Dictionary is "to cause to move, operate or work at the same rate and exactly together." In fact, we are trying to get a speed gear to operate at the same RPM as the mainshaft so we can lock it to the shaft without the hint of a grind in a split-second instance we call a "shift." The synchronizer, or "synchro," in a common performance transmission is made up of a hub, slider (also called a clutch), strut keys, springs, and synchro rings, which are also called blocker rings. The hub is splined to the output shaft. The slider is the sliding component that physically mates the speed gear to the hub, thus locking the speed gear to the mainshaft. In the process of moving the slider to engage the selected gear, strut keys which track with the slider exert pressure on the synchro ring to slow down the gear to allow the slider and gear to couple.

The synchro block pictures show several important aspects of synchro design. These show the difference of how a synchro ring relates to the gear on the upshift as opposed to the downshift. Figure A shows upshift mode. During the upshift, as the slider moves toward the selected gear, the slider's keys exert pressure on the ring, which tries to lock onto that gear. The drag of the ring against the cone of the gear forces the ring to wind up in the same direction the gear is moving. The struts have a two-fold purpose. First is to apply pressure to the ring so that it can lock on the gear and second to "index" the ring. Proper synchro indexing is the ticket to a clean and effortless shift. Figure B shows the direction the ring must take in order for the slider to continue on its path to locking on the gear. Note that the ring must move in the opposite direction in order for the slider to pass through. When you shift you press down on the clutch pedal, which temporarily uncouples the transmission from the engine. Once uncoupled, the gear can slightly "blip" in the opposite direction and complete the shift. It's important to realize that the complete gearset is in constant mesh. As a result, the inertia of all the gears is overcome in order to make a shift, not any one particular gear. During a downshift, the same series of events occurs. The difference is that the power flow starts at the output shaft. Since power flow switches, the synchro rings use the opposite side of the teeth on a downshift as illustrated in both pictures. Sometimes a bad synchro will work on an upshift but grind on a downshift because the ring uses different aspects of itself.

One common misconception is that synchro rings cause jumping out of gear. In reality, synchros prevent grinding in gear. Therefore, if you make a clean shift and cannot keep the transmission in gear, you usually have worn "hard parts," such as a gear, shift fork, or slider. It can be one item or a combination of any. The tiny clutch teeth transmit the power. If they get worn or twisted from lugging a transmission, the spline of the slider applies uneven pressure on the worn tooth and will allow the two parts to slip away. A worn shift fork may not allow the slider and gear to engage fully, and a worn fork groove causes that same problem. Torque locking

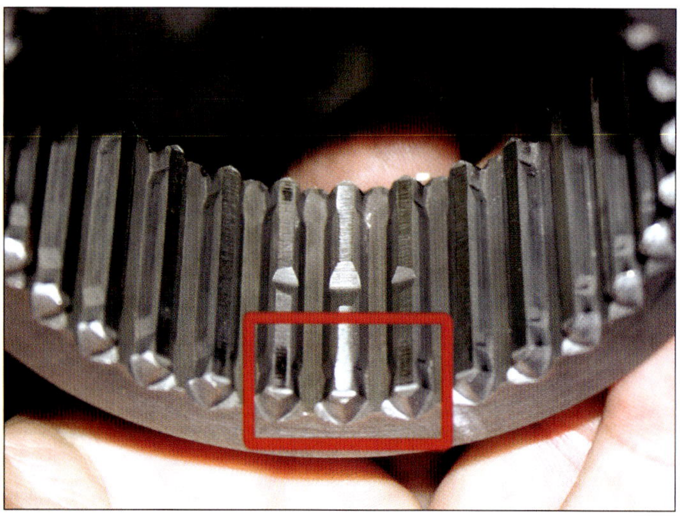

The back cut is referred to as a torque-locking spline, which forces the gear to stay engaged.

A synchro ring that sits correctly on the cone of the gear usually has a clearance of 0.050 inch.

sliders have the ends of their splines back cut. This in effect causes the clutch tooth of the gear to sort of fit in a pocket within the slider's spline, or it forces the gear to ramp in towards the center of the slider whether you're on or off the gas. This design is common in most new transmissions. Some aftermarket gears for older transmissions are adding these features as well.

A synchro ring is nothing more than a cone clutch. The ring has a tapered inside diameter, which sits on a tapered cone of the gear. The ring has tiny threads, which lock onto the cone of the gear. In order for the ring to grab the gear properly, the ring's inside diameter has to be concentric with the mating surface of the gear, and it also has to be smaller. A good ring won't wobble or sit low on the gear. When a transmission grinds going into gear, the synchro ring has lost its ability to stop the gear. The ring is either warped or worn so much that no interference fit is left. I stress this issue because rings are often blamed for jumping out of gear problems or block out problems. This will be discussed in the troubleshooting section of this book.

Gear Design Basics

Today's transmission gears are computer designed to yield the maximum capacity in the smallest package. Fuel efficiency and preventing frictional losses are paramount. Often, people insist on purchasing NOS (new old-stock) gears from the 1960s rather than new products offered by current aftermarket vendors. Somehow, when it comes to old transmission gears, people can't comprehend that in 40 years better methods of manufacturing have developed. The gear sets today can be made with more finely pitched helix angles to reduce noise. In addition, exotic alloys and profiles, which were next to impossible to mass-produce in the 1960s, are now being produced. Heat treating has also reached new levels of accuracy and control.

Early spur-gear designs were very strong but noisy and inefficient. Most 3- and 4-speed transmissions of the muscle car era used helical gears for quiet, efficient operation. General Motors' M22 Muncie 4-speed was the first semi-helical gearbox offered in a passenger car. The gears were at half the normal helix angle, creating a stronger but noisier transmission. This gear noise earned it the "Rockcrusher" nickname. Helical gears, because of their climbing nature, will produce more heat and thrust loads. A spur-gear transmission hardly produces any thrust loads. Most transmissions destined for endurance racing use spur-gear designs because less heat is generated and little thrust loads are put on the bearings.

When BorgWarner released the T5 5-speed in 1982, die-hard manual-shift enthusiasts laughed at the whole design. The T5 was the first American transmission that was truly computer designed. It was never intended for drag racing or road racing. It sported a smaller center-to-center than older muscle car 4-speeds and was designed for light- to medium-duty use. Now, 22 years later, the OEM is still using it, and countless applications have been developed based on that radical design. This transmission started to change people's perception of transmission design, especially when it came to bearings.

Bearing Selection

Some form of a bearing supports upper and lower geartrains in any transmission design. The obvious reason for bearing use is to reduce friction. A properly selected bearing for an application withstands several different loads. The amount of thrust load, radial load, and RPM required for a particular application demands a certain style bearing design. Helical gears produce both thrust and radial loads. Tapered bearings as well as ball bearings can handle both of these loads simultaneously. That is why they are frequently used in transmissions. Needle bearings are often found supporting countergears, but they cannot handle any thrust loading. That is why countergears employ the use of thrust washers. Space can be a deciding factor for bearing selection as well. Tapered bearings, often called cup and cone bearings, can handle the same loads as a ball bearing but in a smaller format. Since tapered bearings are of a two-piece design, they require a preload or some means of keeping the cup and cone together. This is accomplished using a design that requires shims to set the preload. Usually a transmission that strictly uses tapered bearings will be more expensive to produce than one using ball bearings.

There are always gains followed by losses when choosing a bearing. A big bearing may be able to handle a certain capacity issue, but the larger size may require a bigger case or longer center-to-center distance. There has been a debate that you can increase the capacity of a ball bearing by fitting the maximum allowable amount of balls into the raceways. This is true up to a point. A standard ball bearing is called a "Conrad"-style bearing, and the bearing with the maximum amount of balls is called a "Max Capacity" or "Filling Slot" bearing. Conrad-style bearings can have high raceway walls allowing for high-thrust capacities. The filling-slot bearing literally has slots in both the outer and inner raceways. The maximum amount of balls can be dropped in the slots before the bearing cage is riveted into place. The problem is that the filling slot reduces thrust capacity because the raceway walls now have an opening or may not be as high as a Conrad style. However, this added number of balls increases radial load or up and down load.

Bearing design and ball size present some trade-offs as well. If we have a fixed diameter, one can fit more 1/4-inch diameter balls than 5/16-inch diameter balls in the same complement. One bearing may handle loads in one direction better than the other, and one may have more stability than the other. Ball bearings designed for industrial applications may have tighter clearances than the same size bearing used in automotive applications. General automotive grade bearings have what are called a C3 fit. This looser fit enables dirt to pass through the raceways better. Part numbers of bearings are usually etched on the raceways.

All race and street transmissions I build use regular Conrad bearings. Max-capacity bearings develop more heat, which will also kill the bearing's life. I've actually seen more failures with max capacity bearings and generally avoid using them. The fact that a Conrad bearing is less expensive is an added bonus.

The same bearing but different raceways designed for 5/16- and 1/4-inch diameter balls.

By the early 1980s we saw a trend in which transmission manufacturers placed needle bearings under each individual speed gear. Before that, gears rode directly on the shafts, with oil grooves or machined valleys for oil to gather in. It was an inexpensive process that made assembly very easy. Another alternative was sintered iron or bronze bushings. Almost all of the transmissions produced today use needle bearings under all the gears. This produces less drag (peristaltic drag) and allows the transmissions to shift easier. It also aids in better fuel economy. Use of needle bearings under individual gears also prevents seizures at high speeds. Some early T10 and Muncie 4-speeds had rollerized first-speed gears. When used in endurance race situations, the first gear would get hot and weld itself to the mainshaft. Since first gear can be spinning over two times faster than the output shaft it rides on, a car running at 7,000 rpm can come to a pretty nasty stop if first gear overheats.

When I look back and see the huge number of changes in the past three years in manufacturing, one thing sticks in my head. Advances in machine design and gear design software are allowing more and more people to create stronger, more efficient and easier-to-shift transmissions. The world is truly a better place to shift in.

A reduction in thrust capacity decreases the thrust loading of the bearing, which is not desirable.

Excessive heat in a filling slot bearing caused the bearing cage to spread a drop on the inner raceway.

TRANSMISSION ASSEMBLY AND INSPECTION TECHNIQUES

The transmissions outlined in this book were chosen because of their popularity as well as their capacity to be rebuilt with common shop tools. Although you may want to rebuild a transmission not outlined in this book, you must realize that the cost of certain specialty tools required for a one-time job may not be justified. Please take the time to read all of the build-ups. Some of the tips and techniques used on one transmission may be used on another as well, or one not covered in this book. One must also invest in, if at all possible, factory shop manuals. You must also have the ability to read and comprehend why the specialty tools are needed and what their function is. Working on a transmission you have never worked on before without any form of reference is a recipe for disaster. You may find that a required tool's function may be duplicated by another means that you have at your disposal. I remember a Doug Nash overdrive having needed a special spring compressor and retaining-bolt kit. A shop press and three standard 8mm bolts served the same function.

So What Do I Use?

Manual transmissions can be quite old, so if you are going to do a rebuild and not a quick repair, I suggest you purchase a rebuild kit. These kits include all wearable parts, such

Typical kits, like this, sell for under $200 and make a rebuild quite easy. Better companies offer support.

Aluminum bar stock, old bearing races, used wrist pins, or the other pieces you see here help during a transmission rebuild.

as bearings, synchro rings, thrust washers, needle bearings, gaskets, and seals. High-grade kits have OEM bearings and gaskets. Less expensive kits may have "no name" bearings, which are not worth the $30 savings.

As far as tools go, a basic tool kit is a good starting point. Tools that have a lifetime warranty form a basic foundation and are a wise investment. Hydraulic shop presses can be purchased for under $200. They are great for gear work and will prove useful for other projects as well. A few different-size bearing clamps, old transmission cases, piston wrist pins, aluminum round stock, and bearing races are great additions to the collection of press tools.

An assortment of snap ring and retaining ring pliers is always needed. I've purchased pliers at swap meets at rock-bottom prices. Buy up cheap retaining ring pliers so that you can grind them if needed for specialty work. Old drive-shaft yokes will help you align tail housings, and make sure the rear bushings are installed correctly. These also help you spin a transmission. Pry bar sets and screwdrivers, which you don't mind using for anything but screws, are ideal for this work. Two-jaw or three-jaw pullers are nice, but I rarely use them. Propane torches for heating parts help install slip-on components, such as steel speedo gears that could get ruined by pressing or hammering them in place. Telescopic magnets are great for removing detent balls. Cheap dental tool kits are great for removing

Be creative! I find cheap party platters at the Dollar Store and use them as parts organizers.

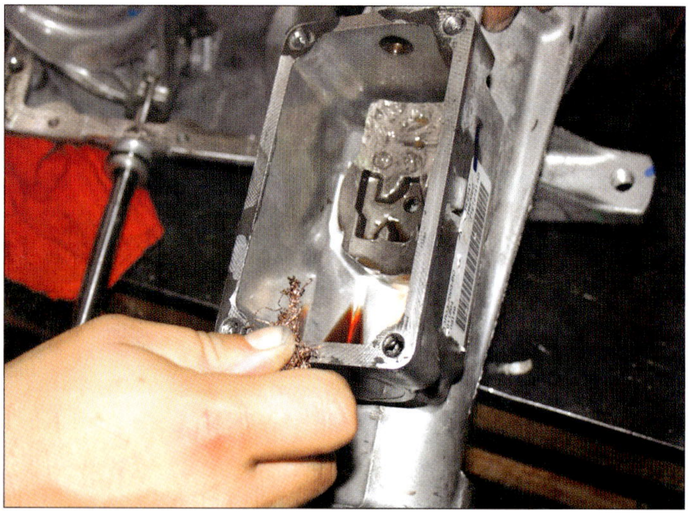

Copper scouring pads are abrasive enough to make removing sealant very easy.

Special chisels used for muffler work make removing tail housing bushings a snap.

Insert the cutting edge of the chisel along the oil feed passage and tap the bushing out.

Use an old driveshaft yoke to make sure your new bushing is pressed in properly.

O-rings and retrieving detent springs. I prefer nonserrated steak knives for removing gaskets rather than special gasket "scrapers." Copper scouring pads are great for removing silicone sealant without marring aluminum surfaces. A big mistake is to use an air-powered rotary scouring pad to remove gaskets. This ruins the precision flat surface of the case and will lead to leaks.

A vise holds some transmissions, but building a custom holding fixture is the way to go if you plan on doing volume work. Personally, I've never used a holding fixture in my 27 years, because setting up a transmission fixture is a time-consuming process. By following the techniques detailed in this book, a fixture simply isn't required. If you are going to clean your own parts, then you should invest in the proper cleaning tank, safety glasses, and gloves. It is easier just to send all of your parts to a production transmission shop and pay to have them pressure cleaned. I also don't have air in my shop. Instead, I use electric impact guns. I hate the noise of compressors, and I don't like blowing off parts because it just creates more airborne contaminants. You can get small air compressors for under $300 that even come with starter air tool kits. In addition, a small can of compressed air typically used for cleaning computers is usually adequate for the one-time job. If you are rebuilding a transmission and you require the installation of new gears, you may need to debur them. A Dremel rotary tool with cut-off wheel is ideal for cleaning up gears.

Bench grinders, drill presses, and a 4-inch disc grinder should be acquired for more serious work such as slick shifting gears or clearancing gears by cutting them back a bit.

Repairing Castings

A transmission doesn't have to fail to have its fair share of problems. The environment and service people can be a hazard as well. I've seen people run over objects, drop transmissions, and get into accidents. This leads to cracks, holes, and stripped threads. Aluminum and cast-iron cases can be heli-arc welded. Welding broken mounting ears, caused by people pulling a transmission into the engine with bolts, is a common job, which should be done by a professional. You can warp a transmission case easily by using too much heat. Epoxies, such as J-B Weld, are great for fixing small cracks or porosity problems. Drilling and tapping out a small hole and then plugging it with a bolt is also an alternative to welding. Alloy main cases tend to stretch when a gear set explodes. Muncie and T10 transmissions have a tendency to leak from the front countershaft bore.

Gear Preparation

Inspect gears for typical cracks, chips, and obvious missing teeth.

Dremel rotary tools have a reinforced cut-off wheel, which makes for very easy deburring.

This front countershaft bore has been bushed. The bushing restores the center-to-center distance. It also creates a larger steel surface area, tighter press fit and prevents leaks. For performance applications, rebushing both front and rear bores and moving the countergear up a few thousands of an inch to help reduce gear deflection.

Gears with missing teeth will not work. Since the gears in most manual-shift transmissions are in constant mesh, if the teeth are broken on one gear, usually the matching gear exhibits some form of damage or stress caused by the upper gear slamming against it. Often, people ask to purchase a new input shaft because the old one is missing three teeth. I'll ask them if they want to purchase a new matching countergear, and the response is usually "no," because that gear is not missing any teeth. These people haven't noticed the bent, cracked, or deformed teeth. As a rule, if one gear has broken teeth, always change the matching gear, too. When transmissions are run low on oil, gears can become blue or black in color. Usually the front gears will starve of oil because oil moves to the back of the transmission when a car accelerates. Gears will appear scuffed looking or gouged out in the center of the tooth. Chipped teeth are another story.

Debris getting lodged in the gear usually causes chipping in gears. Sliding gears that are non-synchronized, such as most reverse gears, will also have chips. Some gears are simply chipped by poor assembly or handling techniques. It's quite common for many new gears in boxes to have their share of dings and chips. Any transmissions I build also have the gears completely deburred. You can debur most chipped gears just as long as the chipped section is no longer than one-quarter the length of the gear tooth and never extends into the root or center section of the gear tooth.

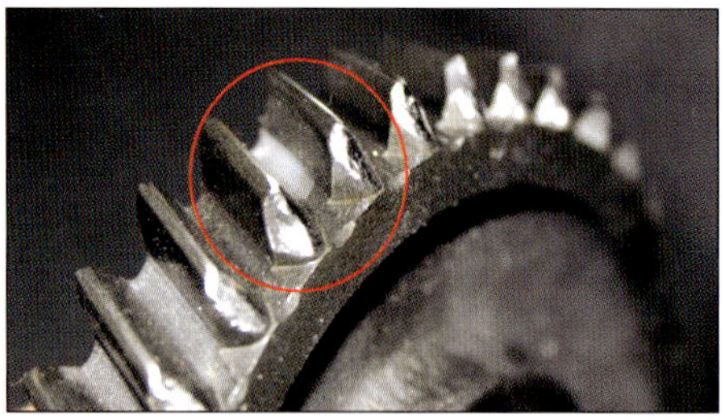

This is common damage on a Muncie reverse gear. While this gear is still usable, the chipped section overhangs the idler gear when engaged.

The gear on the left has badly worn engagement teeth. These teeth are what hold the transmission "in gear." The new gear on the right shows how much material is actually missing.

Anti-rattle or anti-backlash plates are quite common in Muncie, Chrysler, and Saginaw transmissions. These spring-loaded plates are offset in relationship to the actual tooth profile. I cut these plates off because they can loosen up in a performance application.

Bolts and Threads

Older transmissions may have bolts that are seized in place. Using an impact gun is a sure way to either strip or snap the bolt. If a bolt seems hard to remove, try heating it first with a propane torch and strike it with a hammer. You want to hit the bolt flat on the head. The idea is to stretch the threads enough to loosen the bolt. Hex-socket pipe plugs easily strip. To remove difficult bolts, heat and strike them, and then use a hammer to drive a Torx bit into the hex bore. Slamming the Torx bit into the plug deforms the threads and produces a stronger contact pattern to make a stripped hex plug removable. This has always worked for me. If a bolt has broken off, try to get the remaining surface as flat as possible. Use a sharp centering punch in the middle of the bolt. The centering punch will help you start a drill. Drill with a small bit first and work up to the size of the bore so that you can install a threaded insert.

When you have stripped threads, the wrong thing to do is rethread to a larger size. That is unprofessional. Threaded inserts from Heli-Coil make it possible to restore damaged threads to the correct size. The hole is tapped using a special tap that allows a threaded insert to be screwed into the bore. Once the insert is installed, the drive tang is broken off and the thread is restored.

The stripped thread is drilled out to the dimension specified for the new tap.

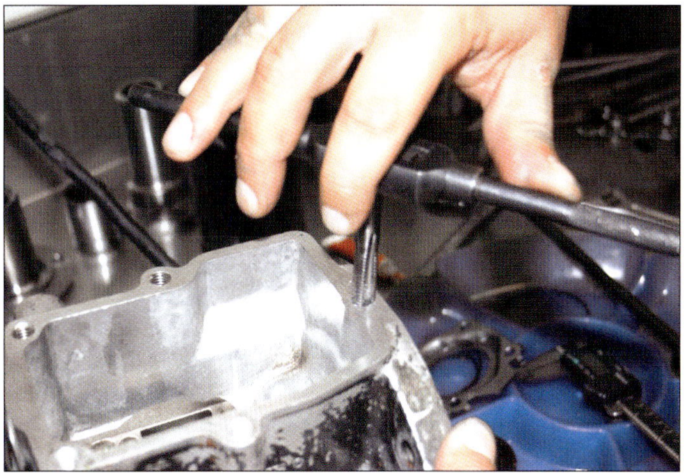

These threaded inserts form standard sized threads from nonstandard threads.

The insertion tool winds the drive tang into place. Once it's seated, the drive tang is snapped off.

Before removing the drive tang, check to see that the new thread actually works.

Hydraulic Press Basics

There is one basic rule to using a press. That is to only press and support the part against the area that is holding the press fit. For example, if a bearing is held in a housing by the outer race, press against that outer race. If the bearing is held on by the inner race, press or support against that inner race. If you support against the inner race and press against the outer race, you will deform the raceways, and basically damage the part. I cut apart older bearings and use the old races as support surfaces. I've purchased cheap socket sets and use the sockets as press tools as well. However, never hammer on hardened parts.

Loading Needle Bearings

People seem to have a hard time loading needle bearings in counter gears. The thought of loading over one hundred bearings in a gear can be intimidating. Oddly, most factory shop manuals advocate the use of "dummy shafts," to load needles. I always say, "dummy shafts are for dummies." The problem with the dummy shaft is that you don't see the needles or the position of the needles in the gear. You can't see if you've put in the proper amount of needles. When removing the dummy shaft and sliding in the real shaft in place, sometimes the needles can drop, which means starting over again. Consider the following procedure and the thought process behind it. I use grease to pack the needles in place. You can use wheel-bearing grease if the transmission is using conventional 75w-90 weight gear lubes. There are new assembly lubes, such as "Trans Gel" or "Trans Goo." These assembly gels break down faster for both gear lube and automatic transmission fluid applications. They also seem to retain viscosity better in colder or warmer climates. If you're working in a 96-degree shop in Florida, wheel-bearing grease will not hold as well as an assembly gel. When loading the needles you can only load as many that will fit. There is no need to count. If you can fit one more needle, then you need that extra needle. Once I have all of my needles loaded, I take the countershaft I'm going to use and push it through the gear. I hold a thrust washer on the opposite end of the gear to keep all the needles from getting pushed out, and then I push the shaft through. This adds some extra packing and squeezes out extra grease. I rotate the shaft in the gear because this allows me to check for binding and smooth rotation. Next, I then remove the shaft, which allows me to visually inspect the needles and check for complete rows, good alignment, and proper spacing. None of these checks can be performed by using a dummy shaft, since the gear is installed in the case after the shaft is removed.

Subassemblies–Synchronizers

Some of the basic assembly techniques pictured in this section are used in just about every build listed in this book. Rather than repeat these steps with each build, they are shown here.

The synchro slider can exhibit wear on the teeth, but can also have an enlarged groove for the shift fork. Common causes for falling out of gear are worn engagement teeth on the slider and gear combined with a worn shift-fork groove and worn shift fork.

In this step, place the spacer tube in the countergear and use plenty of grease.

A generous amount of grease provides ample adhesion, so you can load the needles into place.

Once everything is in place, push the countershaft you are going to use through the gear, but hold a thrust washer in the front to prevent parts coming out of the opposite end.

Inspect the countergear to ensure you've added as many needles as possible and they are all in place.

Synchro rings should be fitted and checked on each gear. Don't assume that a new ring is okay. I've seen many rings with defective machining or warping. You can warp a new ring by dropping it on the floor. A good ring should not wobble on the gear cone. It should fit firmly. It should lock to the gear while applying pressure to the ring as you rotate the gear. Make sure the ring is not sitting too high or too low. The average ring gap is 0.030 to 0.060 inch. A low-fitting ring allows for poor lock up and displaced strut keys. A high-fitting ring can cause binding. A worn gear cone that shows defined ridges will surely prevent a ring from properly grabbing the gear. Sometimes a ring may lock well on one gear but not another. You can also use lapping compound, which is basically an abrasive paste to fit a ring to a gear. Once rings are fitted, the "set" is put aside until the build begins. Always fit rings, and never just place new rings on gears.

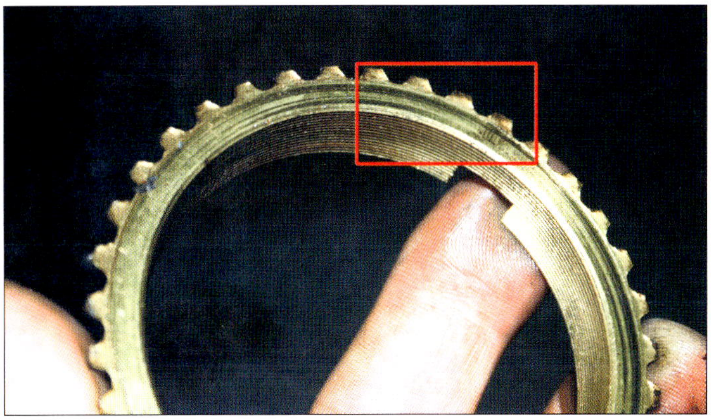

This ring looks good. However, the backside of the ring sits too low and hits the gear.

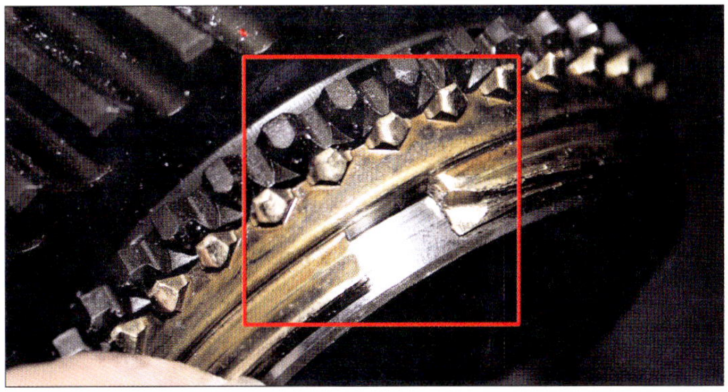

When a ring has ballooned and sits too low, it cannot lock on the gear cone, which causes grinding in gear. As the ring moves away from the strut key, the additional gap allows the strut keys to pop out and further damage the ring. You can see how slipping struts have damaged the key slot.

This shows typically worn teeth on a slider. Once the teeth lose their point definition, they can ruin synchro rings and gears if not changed because the teeth do not effectively mesh.

Notice how this edge is square. It's important to recognize these subtle differences. It can hit a lower countergear if not in the correct position.

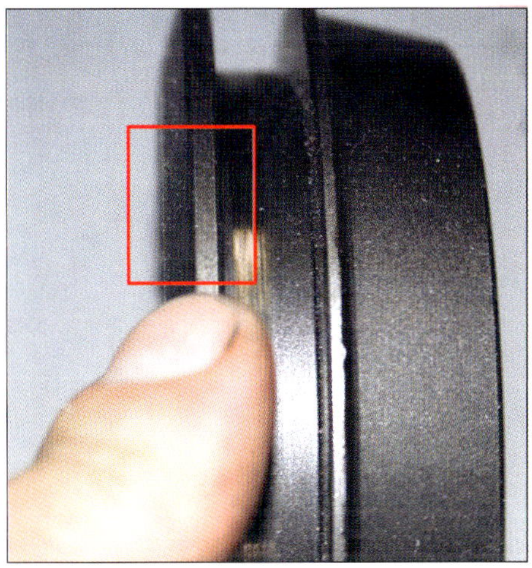

This edge has a taper. This edge will clear the lower gear when the slider is engaged.

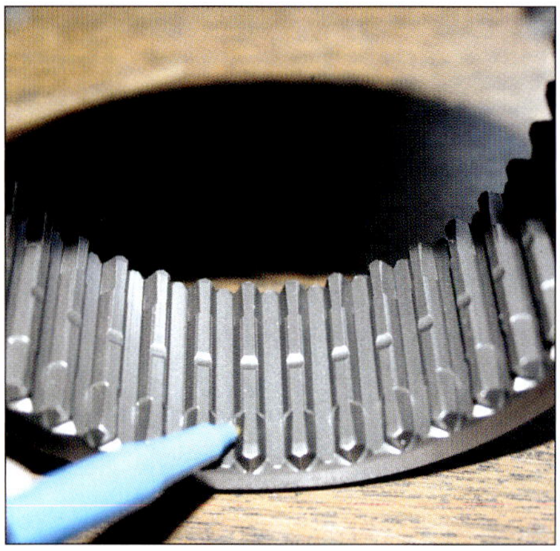

This taper is purposely machined. The back taper or "torque-locking" design forces the gear's engagement teeth to ramp towards the center, so it doesn't fall out of gear.

This is your standard design internal slider spline. It should be used if the matching gears are in pristine condition. If the transmission tends to fall out of gear, the torque-locking sliders are more suited.

Always check the strut keys for wear. If the keys show significant wear, they should be replaced. This key has been worn flat.

Start by placing the slider within the hub and make sure the slider's notches line up with the key slots.

Place all of the struts or dogs in their hub slots.

If using a hollow key, usually the spring tang is placed in the center of the key. Otherwise, it is alongside it. Insert the spring in a clockwise direction.

Flip the synchro assembly over and install the second spring in a clockwise direction. When assembled, the springs will be opposite of one another and create a nice balance.

This completed assembly shows solid-design strut keys. I usually grind the spring tangs so they don't hit the inside wall of the hub. This helps to provide a clean, positive shift.

If the bearing locates from the inside diameter, press or tap it in place from the inside.

Strut keys or shift dogs always give people assembly problems because they are spring loaded. Rather than assemble in the order of first: springs, second: keys and third: slider, which is how most shop manuals tell you how to do it, simply place the slider on the hub first, then the keys and springs last.

This is a modular approach of fitting synchro rings to the gears and creating component subassemblies aids in building a quality transmission. They offer the rebuilder a simple way to perform quality checks. If you have more than one person working in a shop, it is a good idea to have your subassemblies inspected by someone else and vise-versa. Sometimes another set of eyes is all that is needed to prevent a potential disaster.

Bearing Tips

Misdiagnosing bearing noise usually leads to a worthless rebuild. The

If the bearing locates from the outside diameter, press it in or tap it in place from the outside.

It's simple to check snap-ring clearance. See if the edge of the ring will fit the groove before installing it. Then you use a micrometer to measure the width of the groove and thickness of the ring. Dialing in the correct endplay is a matter of subtracting the difference of the two components.

majority of people cannot distinguish between bearing noise and gear noise. Once bearings develop imperfections in the raceways and balls, these imperfections generate noise and vibration. Dust and debris in a bearing can cause noise when spinning it by hand, especially after cleaning a bearing in solvent that has debris in it! Try to get a light into the bearing to inspect the raceways for pits or craters. These usually appear as dark pits. A bad bearing produces sort of a growling erratic noise because the components of the bearing never hit the exact same spot by the next revolution. However, gear noise is never erratic; it's consistent. A broken tooth may translate into a knocking noise that gets faster as the RPM goes up. A whine or hum may change pitch with engine RPM, suggesting worn or mismatched gears. In 4-speed operation in which fourth gear is a direct drive, the input and output shafts are locked and reduce loads on the bearings and gears. This can cause the noise of bad gears or bearings to diminish when in fourth gear. Bearings are relatively cheap. There is no point in reusing 40-year-old bearings when the new bearings of today exhibit better quality standards.

Sometimes transmissions have selective snap rings to set endplay. Often the tightest fitting ring will work, since gear clearances for oil are usually built into the design. Some early T10 4-speeds will actually lock up with a snap ring that is too thick.

Here is a typical sealed bearing. New designs always use them. This bearing had over 100,000 miles and when I pulled the seals the bearing still looked pretty good.

I've taken the sealed bearing apart. If you look carefully you can see how one side of the bearing has more wear than the other due to the nature of thrust loads from the helical cut gears.

Not all bearings of the same dimension are alike. This is something you must detect. The new bearing on the left has a slight chamfer to the inside diameter (ID). The ID of the older inner raceway has a very square cut bore with no chamfer. The bore with no chamfer is a better thrust surface and will support a snap ring better. Better designs use a spacer in front of the raceway to maximize snap ring holding capacity.

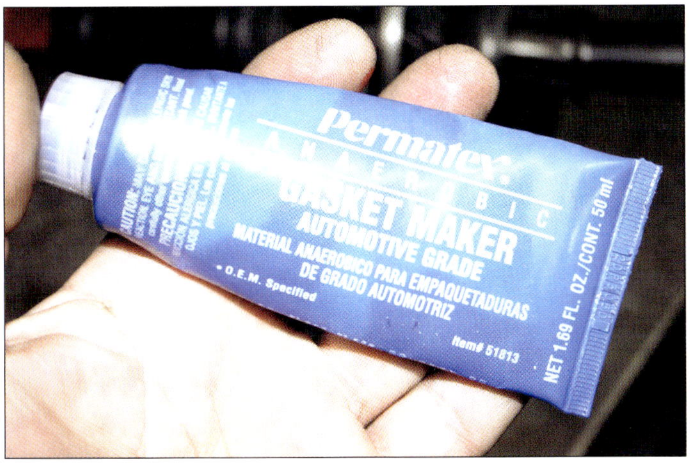

Anaerobic sealants, such as #51813 from Permatex, seal on contact. This is currently used in a great many of OEM applications.

If I can find an equivalent bearing with seals for an application, I will use it. Since dirt is the main cause of bearing failure, seals will keep the dirt out. Oil will actually still get past the seals and keep it lubricated.

Gaskets and Sealants

Good-quality gaskets save you a great deal of headaches. Try to use original-equipment gaskets when possible. There are several companies that now reproduce gaskets using the same compounds the factory did years ago. I always coat both gasket surfaces with either RTV Silicone or anaerobic sealant. I prefer the anaerobic because it sets on contact and cures within 20 minutes. This is great if you need to get something sealed up quickly and back in the car. Silicone sealants don't cure that fast. If a transmission requires a gasket, use one. Using gasket eliminators may be great for an oil pan or valve cover, but sometimes the gaskets in manual transmissions actually set the spacing of the gears as well as endplay. Certain front-bearing retainers will crack when using sealant only, because the gasket fills the gap between the two parts. Without the gasket, you can bend the retainer around the bearing and crack it as you tighten the bolts.

Most lip seals have an outer steel shell with the rubber seal vulcanized onto it. A circular spring surrounds the inside diameter of the seal to maintain its shape as well as contact with the shaft. Before installing any lip seal, grease the seal. Applying a liberal coating of grease to the seal aids in initial lubrication of the seal as well as using the grease to hold the spring in place.

Certain case castings have bolt holes that break into the inside. Most front bearing retainers have bolts that go right into the case. Oil can leak past the threads, so it is important to use some sort of sealant in these areas.

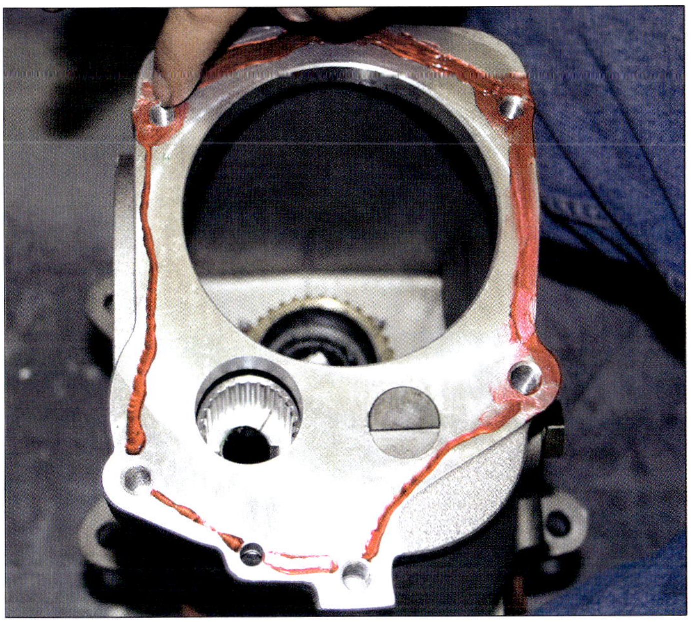

I always coat both sides of the case so the gasket is in-between a skin coat of sealant. It makes for easy removal of the gasket and cleanup.

Grease the internal lip of the seal and spring.

You can use a hammer to drive a seal in place just as long as you do it carefully and avoid distorting the seal. Also coat the outer shell of the seal with sealant, which prevents oil from leaking around the seal due to surface imperfections.

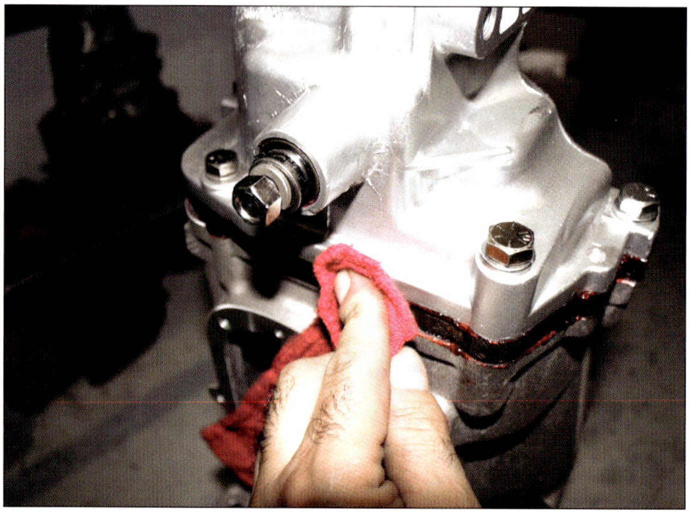

Taking the time to wipe off excess sealant, makes for a better looking rebuild and shows that you are a professional.

Choosing a Premium Rebuild Kit

Manual transmission rebuild kits were rarely available to the general public 30 years ago. Parts were usually obtained through car dealers and parts stores only. The lack of availability of parts forced me to either manufacture or source companies that made the parts for the OEMs. In 1981 I started to sell my own designed kits to the public at 14 car shows a year. We made 12 different kits for popular transmissions. I define a rebuild kit as a kit that replaces all wearable components in a transmission, except for "hard parts." These "wearables" are bearings, synchro rings, gaskets, seals, strut keys, bushings, a countershaft, and small parts, such as needle bearings, snap rings, slingers, thrust washers, and various clips and springs.

A good kit is well thought out and includes everything you need, plus a little extra. Today, unfortunately, people who don't rebuild transmissions are selling kits, or sales people are selling them who really don't know what should or should not be in a kit. An example is the Muncie 4-speed kits, which are now made by many people. A high-quality kit starts with high-quality bearings made in the USA or Japan. Kits having bearings with no name or country of origin on them should not be used. Synchro rings, which are original equipment, should be used whenever possible. The aftermarket T5 cast-brass rings I've seen are a fraction of the strength of the factory-forged bronze alloy rings. Ask if the rings are forged or cast. Cheap kits usually lack a countershaft, which more than likely will need to be replaced since it is a bearing surface. Make sure kits include tail bushings, strut keys, front nuts, and slingers. These are all important items that you will need and are usually left out in cheaper kits.

I have found differences in the fits of front nuts in Muncie kits. Try to get kits with American-made nuts or check the fit. Loose thread will cause the nut to loosen in operation. Good gaskets will usually have a manufacturer name on the paper they are stamped from. No-name gaskets fall into the same category as no-name bearings and should be avoided. A high-quality kit will have bronze or Babbitt-backed thrust washers, just like the OEM. Cheap kits often contain replacement washers made out of stamped steel that will cause premature wear on your gear's thrust surfaces. I have seen insignificant things added to kits to make them more attractive in lieu of quality parts, such as tubes of sealant or cans of paint. Sometimes the kits are padded with cheap sliders or shift forks to appear more equipped, but these still lack quality components. When you do make a purchase, will you be able to get quality support if you need it? That is the most important consideration.

S-A DESIGN

BORGWARNER TRANSMISSIONS

BorgWarner was one of the largest manufacturers of manual-shift transmissions in the United States. Ford and General Motors primarily made their own manual transmissions from the 1950s to the 1970s. Chrysler Transmission Development evolved into a company called New Process Gear. The majority of new process transmissions were fit into Chrysler products, but years later, that company was sold to competitors as well. BorgWarner was different because it had no affiliation with any automobile manufacturer. The company sold its designs to any manufacturer as well as direct to the public via auto parts stores. It also made replacement parts for everyone's transmissions, something only a few companies ever succeeded at. Companies, such as Republic Gear or Perfection Hy-Test, did the same thing, but oddly today none of these companies produce any gears. This is important because you can find variations of certain BorgWarner transmissions in many makes and models.

From right to left: Muncie, ST10, and Chrysler A833 countergears. The A833 is longer because it incorporated the reverse idler as part of the gear. Which one do you think is stronger?

T10 4-Speed History and Facts

BorgWarner produced many 3-speeds, such as the T85, T86, T87, T14, T15, and the T16. The first passenger car 4-speed was the T10. In fact, it was the first American passenger-car 4-speed! It was based on the T85 design. Some T10 part numbers today still use the T85 prefix. The T10 became an option in Corvettes in May 1957. Keep in mind that power levels of engines were climbing, and as a result of increased power levels, the T10 design had to change as well. Gears became wider, and alloys were changed. There wasn't an emphasis on documenting changes. For example, early third-speed gears have a bronze bushing and matching-diameter mainshaft, which do not interchange with later gears and shafts, but in GM versions only. Early T10 transmissions were used in General Motors and Ford Products. As horsepower levels increased, Ford designed the Toploader and GM the Muncie. GM replaced the T10 with the Muncie in mid 1963. Ford kept both the T10 and Toploader in production at the same time. Certain optioned cars received a Toploader while others a T10. I still to this day can't come across specific data that says which Ford cars got which transmission.

To accommodate increasing power levels, BorgWarner designed the Super T10. This transmission featured gears with a wide cut and a coarse pitch for better capacity. It eliminated a thrust bearing between second and third gear to tighten up on endplay issues and ditched alloy main cases in favor of nodular iron cases, which yielded a 50% increase in tensile strength. Ford and AMC used these "First Design" Super T10 transmissions from 1965 to the early to mid 1970s. BorgWarner also implemented its "Power Brute Parts" line. This gave individuals the ability to purchase upgraded components, gears, and complete replacement transmissions from their local speed shop or auto parts store.

These aftermarket transmissions offered special gear ratios. Again, this made the T10 a favorite amongst drag racers since the ratios gave people certain performance advantages over their stock units. The Super T10 gear ratios were designed to improve acceleration. You could get an "ST10" with a 2.42, 2.64, 2.88, and 3.44 first gear.

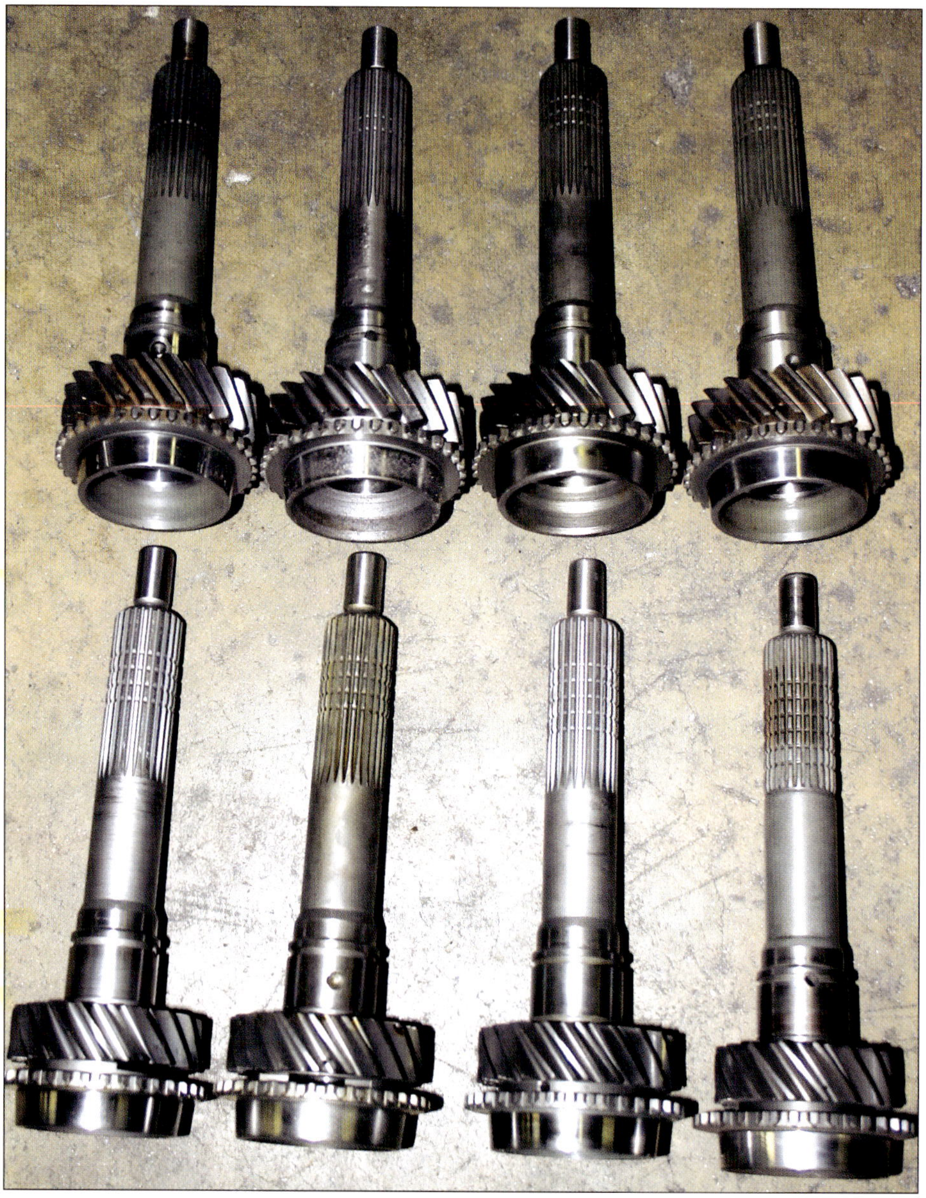

Super T10 GM Input Shafts. Note the various identification grooves and drill marks indicating 9310 alloy.

The 9310 Nickel T10 gears can be identified by drill marks shown on the shaft.

The "Second Design" Super T10 was upgraded in a GM-only version. This GM only upgrade was probably due to the fact that Ford and Chrysler transmissions were much better designed than the GM Muncie 4-speed. In fact, it was quite common to adapt the Chrysler "Hemi" 4-speed to GM applications for drag-race use. The Second Design Super T10 added 26 splines to the input shaft and 32 splines to the output shaft to combat twisting. It increased the countershaft diameter from 7/8 to 1 inch, sleeved first gear (for improved oiling), used a cast-iron side cover to eliminate flex, and eliminated O-ring grooves on the forward shifter shafts to improve strength. The standard alloy ST10 was 8620, but the over-the-counter race units could be ordered with 9310 alloy gears for improved capacity.

The following is a quick ID reference guide for the ST10. Only first gear ratios can be identified by the input shaft ID grooves.

None: 3.42 Ratio – Pontaic
6: 3.42 Ratio – Chevrolet
5: 2.88 Ratio – GM
4: 2.88 Ratio – Aftermarket
3: 2.64 Ratio – All
2: 2.43 Ratio – All

A whole book can be written about T10 identification because many design changes occurred during production. In 1975, when horsepower levels dropped and gas mileage was a concern, GM replaced the Muncie with the ST10. The advantage was the ST10's better first-gear ratios. These worked with more economical axle ratios, such as a 3.08. These Super T10s should not be confused with the replacement over-the-counter transmissions. The ratios and lightweight cases were designed for their intended use, making them weaker than the Muncie 4-speed but better suited to the axle ratios.

The Super T10 was used in GM vehicles from 1975 to 1983 as a 4-speed. The 1982–1983 F-Body Camaro/Firebird had an oddball version with a passenger-mount shifter, to centralize the stick to make it compatible with the 5-speed option. Doug Nash Engineering made a first 2.88:1 gear version coupled to an automatic overdrive for the 1984–1988 Corvette; it was the 4+3. From looking at casting numbers in about 1982, Doug Nash Engineering purchased the rights to the T10 name and design. The design flaws of several components led to seal failures and made the 4+3 a very problematic transmission. By late 1987 Doug Nash Engineering closed and Richmond Gear purchased the rights to the ST10 and all of Doug Nash's designs. The buyout made sense because Richmond Gear was making most of the gears anyway. Today, the GM replacement ST10 4-speed can be purchased new from Richmond Gear, making it the longest running production 4-speed.

BorgWarner Numbering System

With so much misinformation perpetuated by the Internet and online auctions, the BorgWarner numbering system can be an ace in the hole. The early system used a transmission model ID number followed by a part type, followed by a variation. The "A" prefix of T, AT, WT, or AWT was used on all of these ID numbers, more than likely meaning Transmission and Warner Transmission. The "A" in the prefix could stand for "aftermarket," but this is used and mixed with standard factory numbers.

BorgWarner Transmissions

Number	Transmission
10	T10 4-speed
294	New Process A833
296	Ford Toploader
297	Muncie 4-speed
301	Saginaw 3- and 4-speed
302	Saginaw 4-speed
307	New Process A833 (later style)

Part Type	Part Name
1	Main Case
2	Main Shaft
3	Countershaft
6	Front Bearing Retainer
7	Extension Housing
8	Countergear
12	1st Gear
21 & 31	2nd Gear
11	3rd Gear
16	Input Shaft
14	Synchro Ring
83	Synchro Ring
15	Synchro Sleeve
2.5	Synchro Hub
10	Reverse Idler Front and Rear
34	Reverse Idler Rear
23	Shift Fork
148	Side Cover
35	Reverse Idler Shafts
36 & 46	Mainshaft Reverse Gears
40 & 41	Selector Shafts
50	Small Parts Kits

Here's how the early system worked. A countergear for a Ford Toploader is WT296-8. However, a WT296-8A is the same transmission but a different ratio gear. A Muncie 4-speed's first gear is a WT297-12. Since there are three variations of Muncie first-speed gears, there is a WT297-12, WT297-12A, and WT297-12B. There are other less significant part types, such as designators for seals and bushings, but these parts are now in modern rebuild kits.

Current BorgWarner and Tremec Numbering System

While Richmond Gear bought the T10, Tremec Corporation purchased the T5 and T56. Tremec maintains BorgWarner's numbering system on existing production transmissions, including the T5 and T56. New units now have some different numbers that are just starting to evolve as of this writing. Tremec's own transmissions, such as the TKO series, have nothing to do with these systems.

Since 1970, Warner Gear has used a 10-digit numbering system. Old parts that were unchanged kept the old numbers, but if they were revised or improved for any reason, a new 10-digit number was assigned. The first four digits represent the model transmission. The next three define the part category. The final three digits identify the version of that part. A typical number could be 13-52-080-030. Let's break it down.

Transmission Model Designators– The First 4 Digits

ID No.	Transmssion Type
10-00	Miscellaneous Standard Parts
13-00	Miscellaneous Standard Parts
13-04	T10 4-speed
13-32	SR4 4-speed (Ford)
13-38	T50 5-speed
13-40	SR4 4-speed (AMC)
13-50	Ford Ranger Transfer Case
13-51	T4 4-speed
13-52	T5 5-speed
13-53	T4C 4-speed
13-81	T45 5-speed
13-86	T56 6-speed

There are many truck, 3-speed, and transfer-case models that were left off this list. These numbers are also used in casting numbers on main cases, making it easy to identify the type of transmission.

Component Part Category– The Next 3 Digits

I've left out small parts to focus on numbers that a person would actually come across.

ID No.	Transmssion Type
027	Bearing Caps and Retainers
031	Flanges and Yokes
037	Shims
039	Covers
065	Transmission Cases
066	Extension Housings
067	Intermediate Shafts, Torque Carrying
068	Intermediate Shafts, non Torque Carrying
070	Gears Splines or Keyed
077	Countegears
080	Gears, Free Running (1st, 2nd, 3rd, and 5th)
083	3rd Speed Free Running
084	Idler Gears
085	Input Shafts
089	Synchro Sleeves
090	Synchro Hubs
091	Synchro Rings
096	Shift Forks and Lugs
097	Shift Covers
098	Shift Levers
100	Shift Rails
127	Bushings and Wear Pads
171	Mainshafts
671	Mainshaft Assemblies

In the above example of 13-52-080-030, it decodes as a T5 5-speed free-running gear. It's actually a first-speed gear, but unlike the early system, we cannot determine what kind of gear type it is by part number alone.

As also mentioned the system is due for another change. Newer T5 5-speeds manufactured by Tremec now have TTET designators. The T45 5-speed–TNET, T56 6-speed–TNET or TUET. The TKO/3550 5-speed–TCET and the new 3650–TDET.

Rebuilding T10 and Super T10 4-Speeds

Although the T10 and Super T10 may look the same, internal differences make them disassemble in a slightly different fashion. The build techniques are similar to the Muncie, and after a while, one becomes proficient at building this style of transmission. Like the Muncie, they have a side cover, a mid plate-supported mainshaft, reverse shifter shaft in the tailhousing and a countergear supported by a pressed-in countershaft with loose needle roller bearings and thrust washers. The mainshaft and all the gears can be pulled out of the T10 transmission while the countergear and input shaft remain in place. The Super T10 countergear must be dropped, or front bearing removed, first in order for the upper geartrain to clear the countergear during removal.

Early T10 Building Basics

This is your basic early T10. It has a nine-bolt side cover and no external shifter shaft seals.

Remove the nine side cover-bolts, and sidecover. Keep all detents and shifter shafts in the cover. Remove both shift forks.

Remove the four front retainer bolts and front bearing retainer.

Remove maindrive snap ring and spacer.

Some front bearings will slide off while others may have to be pried off.

The reverse shifter shaft lock pin must be punched out from the bottom. This is a tapered pin and can ruin the case if removed in the wrong direction.

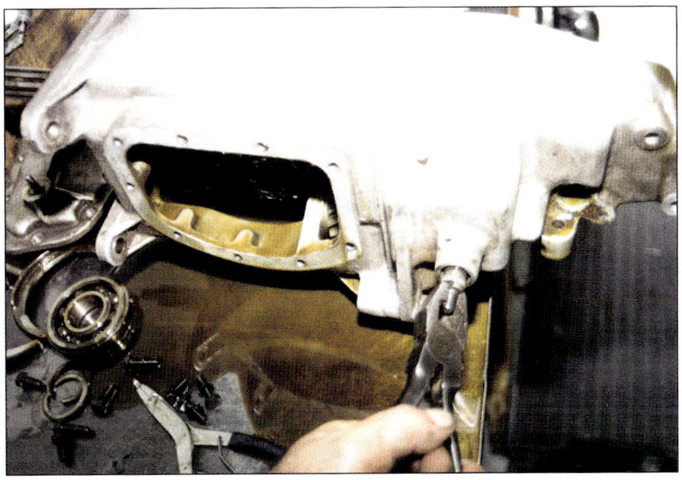

Pull the shifter shaft out and then disengage the reverse fork from the reverse gear.

Once the tailhousing bolts are removed, you can pull the tailhousing off. The rear reverse idler gear may fall out.

Pull the mainshaft assembly out by prying on the mid plate. Some versions have a retaining bolt that has to be removed first. There are only two snap rings that hold the whole gear train together. Remove the 3-4 synchro snap ring first.

Once the 3-4 snap ring is removed, you can slide off the 3-4 synchro assembly, third gear, 2-3 thrust washer, and second gear.

I have disengaged the rear-bearing snap ring from the mainshaft. As I support the remaining parts on the 1-2 synchro, I can press the mainshaft through everything. This includes the pressed-on speedo gear as well.

Here is a top view of how you can simply disassemble the mainshaft in two sections.

Remove the rear bearing by spreading the snap ring open. Sometimes the bearing will drop out; other times a hammer is required.

Remove the input shaft and the forward-reverse idler gear and thrust from the main case, and pick out any loose needles and parts.

Punch the countershaft out of the case from the front towards the back. This will also knock out the woodruff key located in the rear of the case.

Remove the countergear and thrust washers from the main case. New needles and spacers are in most rebuild kits, including a countershaft. The spacer tube is not included, so don't lose it.

The reverse fork is the only part you need to remove in the tail. I leave the reverse shift shaft in the tail. The reverse-shift shaft can be pushed towards the outside to expose the O-ring seal, so it can be changed without having to disengage the detent ball and spring.

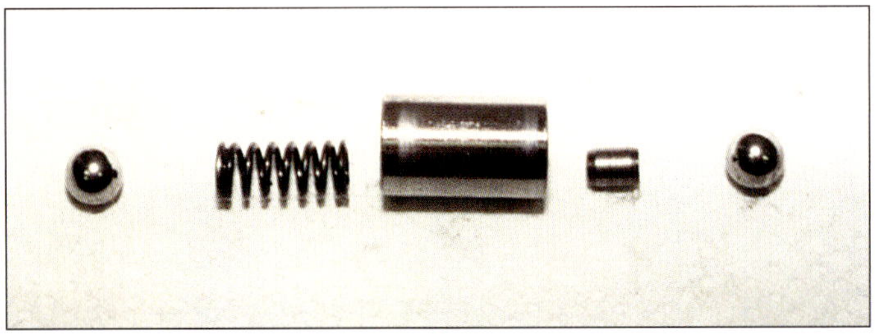

Here is how the interlock system stacks in all T10 and Super T10 4-speeds. The balls, spring, and small pin are your detents. The large tube they fit in is actually your interlock.

The last step of the disassembly and first step of the assembly is the side cover. Remove the shifter shafts, change the O-rings, clean the cover, and put it back together. This keeps the small detents and interlock from getting lost.

Start your assembly by placing the 1-2 synchro assembly less bronze rings on the mainshaft and the first-speed gear sleeve.

We basically reverse the assembly process. We add the new first-gear synchro ring, first gear, rear bearing retainer with new bearing, and replace the snap ring. Rebuild kits have selective fit rings, and we fit the largest ring we can in the groove. We then add reverse gear and the speedometer drive gear. That will complete the back portion.

T10s have a common second-gear jump-out problem. This one had a new second gear and slider installed. We add the second gear ring, second gear, 2-3 thrust washer, third gear, third-gear synchro ring, and the 3-4 synchro assembly.

Using assembly lube, I've held the reverse idler thrust washer and reverse fork in place and shifted it into the forward position, so it's easier to connect it with the reverse gear. Notice I've lubed the reverse shifter shaft O-ring as well. The O-ring is replaced by pulling the shifter shaft outwards, rather than removing it and having to deal with assembling the detent ball and spring.

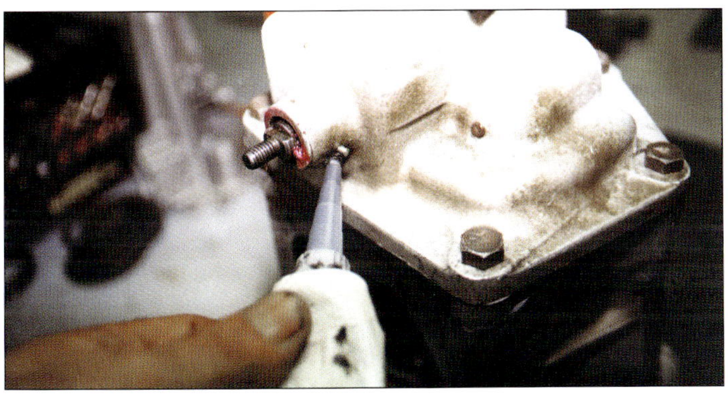

Early T10s use a hollow stamped strut key. The tang of the spring fits inside the key as shown. The spring on the opposite side has its tang located in the same key, but it loads in the opposite direction to give proper balance. Use this method for both synchro assemblies. This is the 3-4 synchro assembly. Use a selective snap ring that gives you endplay between third and second gear between 0.010 and 0.030 inch. Try and keep towards the tighter spec.

Lay the case face down with the countergear installed, input shaft in place without the front bearing but with the needles loaded, the forward reverse idler, and thrust washer.

Reverse the procedure for installing the mainshaft assembly and tail housing. The procedures are similar to the Muncie and Super T10 4-speeds. Apply assembly lube on reverse shifter shaft O-ring before tapping it towards reverse gear. Install tail bolts, speedo fitting, and make sure to put a drop of sealant on the reverse-shifter shaft tapered lock pin.

Install the new front bearing, snap-ring washer, and fit the largest snap ring that you can in the snap ring groove.

Install the front bearing retainer and make sure you apply sealant to the four bolts. GM T10s have no front seal, while Fords and AMCs do. So make sure to replace the seal as well.

It is a good idea to replace the shift forks with new ones. New forks help ensure the transmission does not fall out of gear. The transmission is shifted into second gear before the sidecover is installed. Install the remaining 3-4 shift fork and side cover and you are done.

Take off the front bearing retainer and the side cover, and remove the shift forks. Before removing the tailhousing, punch the tapered pin from the bottom up. Once removed, pull the reverse shifter shaft outwards to disengage the reverse fork from the reverse gears. Unbolt and remove the tail.

Rebuilding Super T10 4-Speeds

As mentioned before, the Super T10 or ST10 disassembles are similar to the T10 and Muncie 4-speed. The main difference is the sequence for removing the mainshaft and gears without using special pullers. Muncie and T10 mainshafts with all the gears intact can simply be pulled out. However, the ST10 has an upper third gear that cannot clear the lower countergear, but there are two ways around this problem. The first option is to remove the front bearing, which allows you to raise the mainshaft while attached to the input shaft over the countergear and take the whole upper assembly out. This technique requires a special puller to yank the front bearing off. The second option is easier: Remove the countershaft and allow the countergear to drop down and away from the upper mainshaft. Once the countergear is dropped, remove the mainshaft and press the input shaft out through the front bearing using the case as a support. With some ratios, you are able to remove the input shaft with the front bearing attached from the inside, allowing you to firmly press it off using a hydraulic press.

I rarely remove the reverse idler shaft because it seldom shows any sign of wear. If you need to remove this shaft, punch the retaining pin cover and pin towards the shaft.

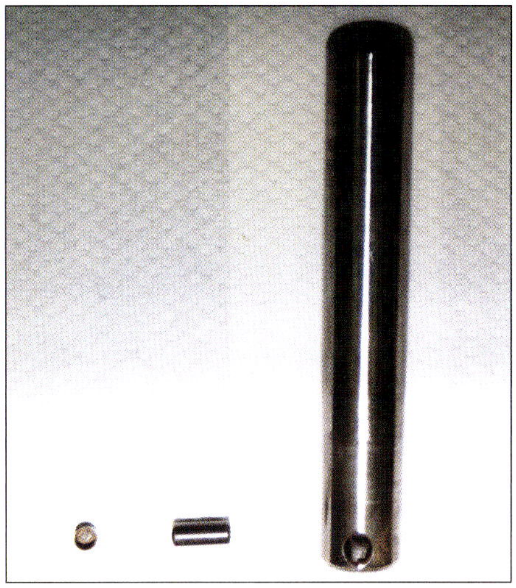

This is what the shaft, pin, and pin cover look like when removed. To replace the shaft, simply insert the shaft back in the tail and punch the pin and pin cover in place from the outside. You should apply sealant to the pin bore.

With the tail and reverse idler removed, unbolt the mid-plate retaining bolt and punch the mid plate's locating dowel into the case and remove it. Once the dowel is removed, you can rotate the plate to expose the countershaft, so the shaft can be removed. Once the shaft has been removed, the whole gear train will have the clearance to come out.

Once the input shaft has been removed, use an old wrist pin as a press arbor and press off the front bearing. I'm using an old case to support the gear.

It is not necessary to remove the reverse shifter shaft. Push it all the way out, change the O-ring, and carefully push it back in a little.

The basic components of the mainshaft all disassembled.

Remove the 3-4 synchro snap ring and remove the third gear and the 3-4 synchro assembly. These are usually slip fit and can be gently tapped off.

Remove the rear bearing by spreading the retaining ring and dislodging it with a good old hammer.

Remove all snap rings, speedometer drive gear and reverse gear from the mainshaft. With the rear bearing snap ring removed, you can press second gear, the 1-2 synchro assembly, first gear, first-gear sleeve, and the rear-bearing retainer in one step.

The countershaft usually fits tightly in the Super T10, and I recommend using a press to remove it, since driving it with a hammer and drift can spread the end. The countergear is loaded like the Muncie. It has a spacer tube, four rows of needle bearings, and six spacer rings. You start loading the countergear with the spacer tube, spacer ring, needle row, spacer ring, needle row, and then spacer ring. Flip the gear around and add a spacer ring, needle row, spacer ring, needle row, and sixth spacer ring. The outside rings should be recessed in the gear slightly.

Pre-fit your countershaft in the countergear to check for binding and rough spots. This procedure also aids in packing the needles.

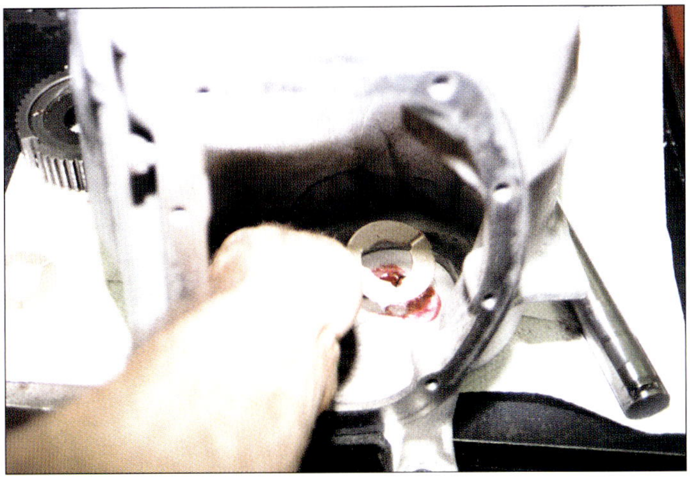

Start reassembly by installing the forward countergear thrust washer and holding it in place with assembly lube.

Once the countergear is in place, slide in the rear thrust washer, engaging its tab in the slot in the case.

You can sink the countershaft far enough down by hand to make sure all the thrust washers line up. I press it back in place and stop about 3/8 inch before the final press to add the woodruff key. Then sink it home.

With the main case facing down, insert the countergear. I'm holding the front spacer ring with my index finger, which keeps the needles from sliding out.

Here is what the input shaft needles look like once loaded in place.

Flip the assembly around and press on the first-gear bushing. Some will slide on, but most require a mild press.

I drop the input shaft in place without the front bearing. The whole front case subassembly is facing down, and an old front case supports it. Add the forward reverse idler and its thrust washer and the fourth-gear synchro ring.

Start assembling the mainshaft by sliding on second gear and the 1-2 assembly. Notice the direction of the slider. The hub's protruding edge is facing second gear.

Flip the shaft around and install the first-gear synchro ring, first gear, and first-gear thrust washer.

The rear-bearing retainer with a new rear bearing sometimes drop in place. I tap it down with a punch if it seems snug, and this is how I do it.

Make sure the rear-bearing 0.062-inch-thick snap ring washer is installed first, and then fit in the thickest selective snap ring in your rebuild kit.

Install the reverse sliding gear, forward speedo gear snap ring, speedo drive gear, and rear speedo gear snap ring. These two snap rings are usually 0.087-inch thick.

Install third gear, third-gear synchro ring, 3-4 synchro assembly, and its snap ring. Usually this ring is 0.087- to 0.091-inch thick. I always use a 0.087-inch one here. The last part is the needle-bearing thrust spacer. I hold this in place with grease. Notice the direction of the slider and hub.

I assemble the ST10 with the main case on top of an old main case. I apply a small bead of sealant to hold the gasket in place and form a perfect seal.

Before installing the mainshaft assembly, pull the 3-4 slider forward like this. Be careful not to let the strut keys fly out.

This picture shows why the 3-4 hub is pulled forward to clear the countergear. I've reinstalled the dowel in the mid plate and will carefully mate the mainshaft to the input shaft. Make sure you line the fourth-gear synchro ring key slots with the strut keys before tapping the whole assembly in place.

Once the mid plate is sitting flush, install its retaining bolt.

Install the rear reverse idler gear. You really have to use a combination of RTV sealant and gaskets on this transmission because the two lower bolts never enter the main case. This area can warp since it is not supported by the main case.

I've installed the reverse idler shaft and thrust washer, and I am inserting the reverse fork and holding everything in place with assembly lube. Note that the reverse shifter shaft is moved forward in the reverse position.

Once the tail is flush with the mid plate you can bolt it down. I clean off excess sealant to give the build a "professional" appearance. Remember to install a new tapered pin, which comes in our rebuild kits.

By applying pressure with my thumb, I'm holding up the reverse gear with my fingers and holding the reverse fork against the reverse gear. You catch the fork onto the reverse gear and gently lower the tail in place.

Once the bearing is in place, pull it out and install its outer-case retaining ring. Install the 0.062-inch bearing snap-ring spacer and then fit the thickest snap ring in front of it.

The front bearing can be installed the same way the mid plate was installed. Use a punch and work the bearing down while holding the input.

Some bearing retainers require thicker gaskets. Our gasket sets include two front retainer gaskets. Place the bearing retainer on the transmission and use one gasket as a "feeler gauge." If you can pull the gasket out while holding the retainer against the bearing, the retainer won't seal and another gasket is required. Obviously, change the front seal, use sealant on the four retainer bolts, and bolt it up.

After the detents and interlock are installed, check for smooth motion. You may want to bolt on the linkage arms so that the mechanism can't fly apart. This is what it looks like in neutral from the inside.

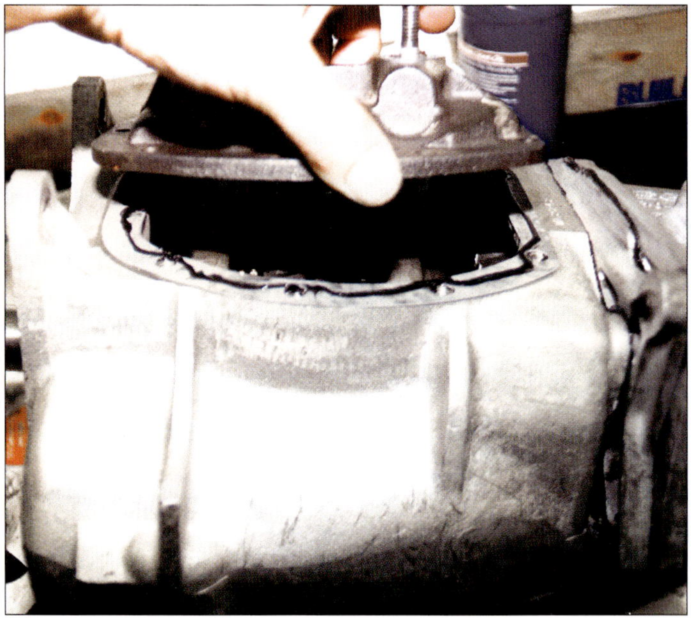

I've installed a bead of sealant on both sides of the side cover gasket. The shift forks are lying on the sliders. You will have to put the transmission into second gear to get the 1-2 shift fork in place. You can leave it in second gear, but you must remember to shift the side cover into second as well. Once in place, install the nine side-cover bolts, and you are done.

More than likely the side cover will come apart when removing it. Install the shifter-shaft seals first, and then install the shifter shafts with care to avoid pinching or rolling over the seals. Notice the sequence of the components in the picture. The outer sleeve is actually the interlock mechanism. The spring, balls, and pin are the detent mechanism.

T5 5-speed History and Facts

A T5 Primer

In 1985, I was doing a great deal of rebuilding for several new car dealers. A T5 transmission had come in with a broken third-gear section from a local Ford dealer. It was different from the normal T5s I had been used to working on. The first thing I noticed is that it no longer had bronze synchro rings but ones made of a fiber material similar to a clutch lining in an automatic. After doing some research and making a few calls, I found out that the T5 was upgraded to "world-class" standards.

There is a reason why I place this little bit of information first. It is important to understand that Borg-Warner, the manufacturer of the T5, was notorious for making unannounced and undocumented changes to production transmissions. T10 4-speeds, SR4 4-speeds, T50 5-speeds, and T4 and T5 transmissions all had a variety of design changes during a production run. Undocumented changes can cause service nightmares and the generation of misinformation. These changes are usually for the better. As certain designs fail when put into a production environment, new improvements are created to remedy warranty issues.

The T5 evolved from the T4, which evolved from the SR4 4-speed. The SR4/T5 concept was a departure from conventional T50 5-speed design. The SR4 stood for "Single Rail 4-speed." The T50 5-speed used three rails. In fact, BorgWarner seemed to be the only company continuing to design single-rail shifting transmissions. Tremec did make the SROD (Single Rail Over Drive), but this design had many flaws, and new units went back to the three-rail system. When I researched the patents on the T5, it was interesting to find out that patents were approved for the fiber-lined synchro rings and single-rail shifting mechanisms in 1981.

In a three-rail system, the shifter is connected to a shift rail, which selects three other rails, such as the 1-2, 3-4, and 5-reverse rails. In a single-rail system, one rail rotates and by use of shift lugs shifts individual 1-2, 3-4, 5-reverse synchros. The single-rail design, if implemented correctly, produces a lighter shifting mechanism as well as a very positive shift movement and feel. By using lightweight components, such as powdered metal interlocks and aluminum shift forks, shifting is much quicker.

The early non-world-class T5s used a combination of flat roller bearings and thrust washers on the countergear. The input and output shafts used tapered-cup and cone bearings. This design seemed to be influenced by the SR4 and T50, which both used a combination of flat roller bearings and thrust washers. By nature of design, flat roller/thrust washer systems cannot allow for preload. This allows gears to move around and get out of mesh. Non-world-class boxes also used all bronze synchro rings. The world-class version allows for preload since it uses tapered bearings on the countergear as well. The drawback is that you have to use shims to set endplay or preload, and that requires more time. The world-class T5 uses fiber composite synchro rings on all gears except fifth. This is similar to material used automatic transmission friction plates. All speed gears in world-class boxes ride on needle bearings except for fifth gear. This concept produces less parasitic drag and improves shift response. The use of fiber synchro rings and tighter tolerances requires lighter weight oils, such as Dextron III. Use of conventional gear lubes will ruin these transmissions.

The world-class spec did not replace the non-world-class design or NWC. In fact, both designs were produced at the same time. It boils down to what money manufacturers wanted to spend. T5s were used in General Motors, Ford, AMC, TVR, Cosworth, Izuzu, Australian Holdens, Korean SSANGS, Nissan, Nisson, Panther Cars as well as Jacobson tractors and several utility vehicles. Beijing Warner in China also manufactured the transmission for Beijing Jeep. As of this writing, there are over 260 T5 assembly part numbers.

T5 identification questions are common. It ranks number one in my popularity poll. So far, the basic T5 has been produced for 26 years, from 1982 to 2007. With so many variations available, it has become very popular for a multitude of projects and conversions.

Rebuilding the T5

There are several variations of the T5. Besides having the world-class and non-world-class specs, there are subtle changes in the world-class spec itself. The buildups cover world-class transmissions because of their current popularity. You may have heard of "Cobra"-style T5s. These have a tapered bearing in the input shaft as opposed to flat rollers. Some also have a reverse brake, which acts as synchro for reverse gear. T5s do not use any gaskets. The factory used black RTV, but I recommend Permatex 51813. A rebuild kit is a must for a T5 because

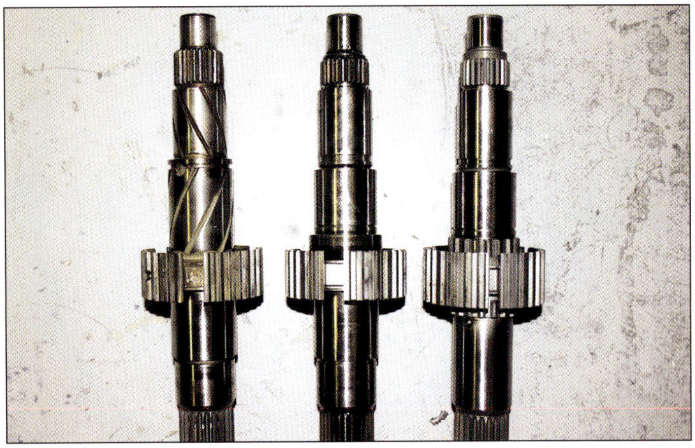

From the left, a T5 NWC mainshaft, a hybrid NWC mainshaft with roller bearing surface, and a WC mainshaft are shown.

The NWC bronze ring to the left and a lined WC ring to the right. Both are 3-4 rings.

individual components tend to cost as much as a whole kit. Variations are covered in the build up.

Now that the T5 is more than 20 years old, some units may have already been rebuilt or assembled with a variety of parts. With over 265 T5 models, you can mix-and-match cases and gear sets to a point. If colored sealants (red, blue, or clear) have been used, it's tell-tale sign that the T5 has been taken apart.

It is not necessary to completely disassemble a T5. The top cover need not be disassembled and lower countergear need not to be removed unless these components in the cover, the countergear, or its bearings need replacement.

There are several procedures that you may or may not have to do. The first optional procedure is to remove the reverse idler gear and shaft, which are simply held in with a roll pin. The other is removing the front countergear bearing or front input shaft bearing since these bearings actually never fail. If the countergear isn't broken, you may want to leave the whole gear still in the case and avoid that step altogether, if for example, you are just changing synchro rings.

The T5 transmission has many variations and many are exported to China, Korea, and Europe for use in foreign vehicles.

While many people assume the T5 only came in Ford and GM variations, it was actually offered in many different spline configurations. The photo shows a large selection of T5 input shafts.

T5 Assembly–Starting from the Ground Up

All T5s we build have all of the gears deburred. If they are used for heavy-duty street/strip use, we add one of our countergear stabilizer plates. We designed the plate to keep preloads constant as well as add steel support for the rear-countergear bearing race. Aluminum copies really don't accomplish anything and are a waste of money. The T5 requires only two measurements for assembly. We've found that if you preload the countergear with a load of 4- to 6-inch pounds of drag, and you keep the upper geartrain to more than 0.003-inch to zero endplay, you will extend the life of your T5 compared to factory specs, which call out for end play.

High-Performance T5

In 1995 I thought the T5 would make a great road-racing transmission for cars up to 400 hp. To me there were plenty of transmissions rated for over 500 hp, and I felt the use of these *big* transmissions in 200-hp application would simply rob too much power. The JT5 was created for a Jaguar race team in late 1995. *(continued on page 66)*

These are the very basic tools needed for a T5 build up. Access to a hydraulic press and a bearing clamp is necessary.

Remove the four shifter-cover bolts and pry off the shifter. Jeep units have five bolts.

Too much sealant can get lodged in the housing bolt bores. When you screw in the bolts, hydraulic action will blow the case apart, especially if you use air tools.

Punching the offset lever dowel pin down disengages the lever from the rail.

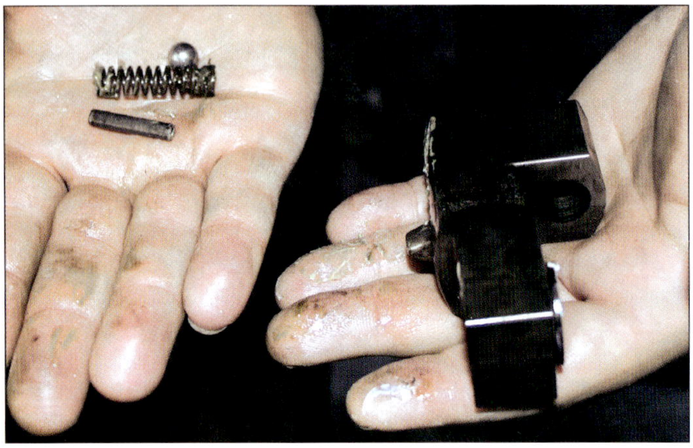

I always make sure the pin is completely removed from the lever, and all components are stored in my parts tray.

Remove front retainer bolts and pry retainer off. There are special pry areas on the case so you don't damage sealing surfaces.

Remove the tail's bolts, pry the tail off, and remove the tail and offset lever in one shot. Be careful not to lose the detent ball and spring.

Remove all the cover bolts. Pry on the cover to break the sealant bond.

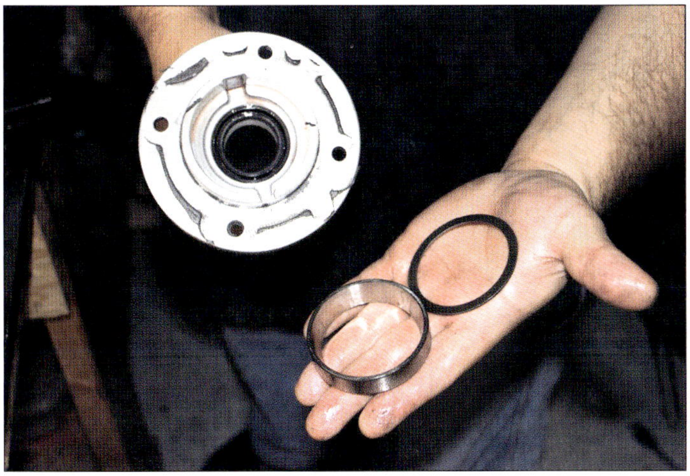

Inside the retainer, you will find the front bearing race and shims. Make sure you take a thickness measurement. If the shims have not been damaged, the same shim can be reused. Also, the original shim size makes a good starting point when starting fresh with new gears.

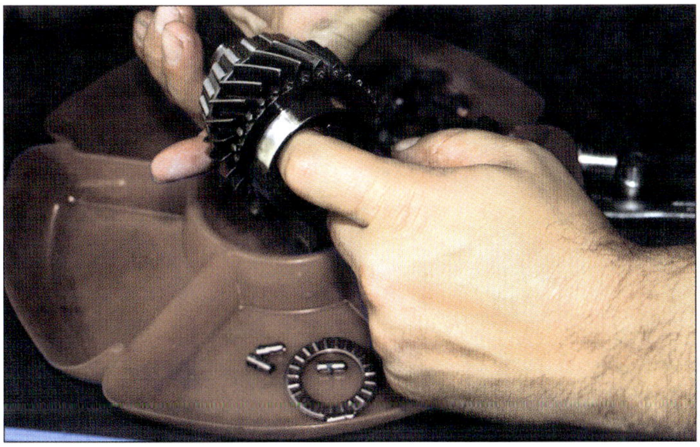

I remove all loose needles and the flat roller and store them in my party platter. Some needles may fall out into the case.

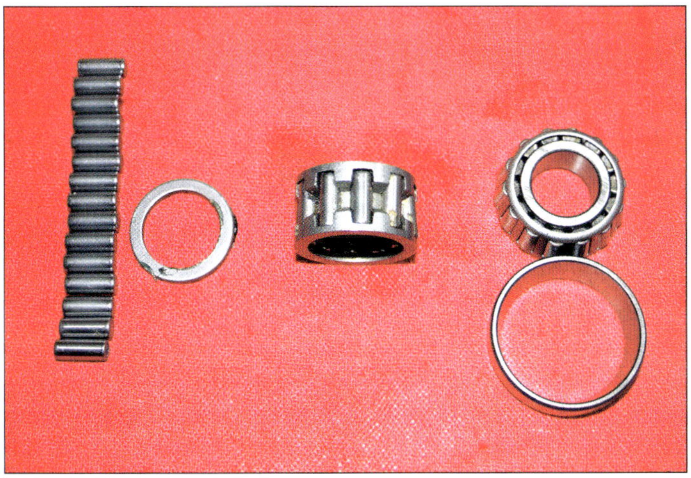

The input is removed by pulling it out of the front of the case. Rotate and pull at the same time to clear the lower countergear. Also extract the fourth-gear synchro ring and any flat thrust washers.

Support the fifth-gear fork with a socket and drive out the fork's roll pin.

The three types of main drive needles in T5 transmissions. From the left, 15 needles and a spacer, a caged roller found in most European T5s, a tapered cup, and cone bearing in later Cobra-style Ford T5s.

Flip the transmission on its face. This makes it easy to remove the fifth-speed assembly snap ring and pry off the assembly and fork.

With the fifth-gear fork and synchro assembly out (1), I can remove the lower fifth gear (2), and rear mainshaft bearing race (3).

The gear train will lift out of the top. Notice that you don't need to mess with the fifth-reverse lever.

The left fifth-speed synchro retainer is a standard T5 type. The one on the right is the reverse brake style.

Variation! 0.80 inch OD sets and aftermarket fifth gears are larger than the rear bearing race. You can't remove the rear race and mainshaft. It is next to impossible to remove upper fifth gears with a puller, so I drop the countergear, push the upper gear train in the case, and cut the bearing race in to spots 180 degrees apart. This allows me to gain the clearance needed to remove the mainshaft once the rear race is gone.

Press on the clip to release the speedometer drive gear. Upper right corner shows later-model speedometer reluctors, which also act as a secondary retainer for fifth gear. You must remove the reluctor first before removing the fifth-speed snap ring.

Never try pulling fifth gear off or using press clamps against it. Use press plates and support under first gear where you get optimum support, and press first through the upper fifth off.

All these parts will come off in one action using a press.

Flip the gear train around, support with press clamps under third gear, and press the mainshaft through the 3-4 synchro and third gear. Units with the Cobra pocket bearing require you to press under the bearing first, remove the bearing, then remove the 3-4 synchro snap ring before doing this step.

Here are all the parts that come off the mainshaft when the previous step has been performed.

Laying parts out on a bench in the order in which they were removed helps you to reassemble it more easily.

Remove the second gear-retaining ring, thrust washer, second gear, bearing, and bearing spacer.

Use a small screwdriver to start unwrapping the spiral lock retaining ring that holds the second-gear thrust washer and synchro ring in place.

Sometimes the 1-2 slider will fall off from the first gear side. It is the final part to slide off; don't lose the keys and springs. My finger is pointing to a small pin, which keeps the first-gear sleeve from spinning. Don't lose that either. Later units use a tiny ball bearing. So be careful.

Remove the four countergear retaining plate bolts, the plate, and shim. Use your press clamps to clamp under the rear bearing. This will elevate the countergear just enough to press it through the rear bearing.

Once the rear bearing is removed, the countergear comes out of the main case. Again, notice that I never removed the fifth-reverse lever.

There is no way to remove the front countergear bearing without destroying it. So I break off the cage of the bearing, which gives me a solid lip to fasten my press clamps on.

I support the gear on an old transmission case and press it through the front bearing race. Now both bearings are removed from the countergear. Your T5 is now completely apart.

Starting with the bare case, I coat the front countergear bearing bore with Permatex 51813 sealant. The fifth-reverse shift rail Welch plug does not need to be removed. However, if it was removed, now is a good time to reinstall it.

Install the front bearing cup carefully. Tap it in gently and avoid ripping the O-ring.

Install the fifth-reverse shift arm and clip. The pin that it rides on doesn't have to be removed, but if you do, reinstall it with red liquid thread locker.

For clarity I removed the arm, but installed the reverse idler gear, shaft and O-ring, and pinned back in place. Install the roll pin so that it is centered in the shaft.

Press on the front countergear bearing using a simple socket and hammer. Remember to place the rear bearing on it after *it is installed in the main case.*

Simply tap the bearing on with a small punch slowly, and work evenly from side to side. Yes, a long pipe would work as well, but for the one-time rebuild, this works fine.

The left shows the standard retainer plate, and the right shows our beefed-up stabilizer. I have found that under heavy load, the factory plate stretches and that allows for the countergear to move around too much. It is not needed for a standard rebuild.

I'm using a 0.132-inch-thick shim here. Lately I seem to average 0.125 to 0.135 inch. Our new peel-'n-place shims use a solid 0.080-inch shim and a peelable 0.080-inch shim in 0.005-inch increments.

Whatever plate you decide to use, shim it to yield 4- to 6-inch pounds of drag. If you're doing it by "feel," the countergear should have no end play and some smooth drag to it.

"A" variation: Insert reverse idler fork, return spring and insert rail. Once in place, turn the rail so the roller engages the fork. "B" variation: This should be done before any gears are in case. Insert wrap spring on fork. Slip spring through reverse arm and fasten to case post. Once in place, insert fifth-reverse arm Torx bit shaft. Then insert reverse idler gear assembly and finally fifth-reverse shift rail.

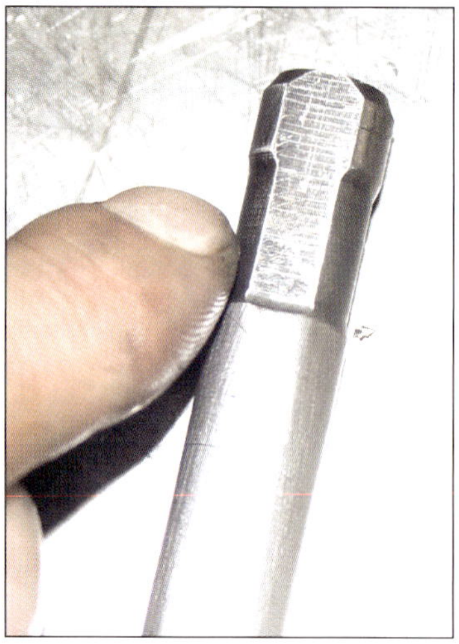

Debur and polish all shift rails. These rails have a tendency to get chewed up from dirt in the case bores.

Later units have specific locations for strut keys. Make sure the central key slot lines up with the key slot in the synchro hub. This applies to the 1-2 as well as 3-4 synchro assemblies. Install the 1-2 slider with the fork groove facing towards the front and install the three strut keys and two springs.

Pre-lube the synchro rings in Dextron III fluid prior to installation. Install the second-gear ring set first.

After installing the second gear synchro ring, install the second gear thrust washer, and spiral lock retaining ring. A small screwdriver can do this quite well.

Install the second gear first, *the bearing spacer, and needle bearing. This method prevents the tiny spacer from getting caught under the gear and bent.*

Install the 2-3 thrust washer and its snap ring.

Install third gear and its synchro ring first, *then the third-gear needle spacer and needle bearing.*

Install the preassembled 3-4 synchro assembly. I use an old socket and smack it down. If building a Cobra-style T5, install the 3-4 hub snap ring and press on the pocket bearing.

Flip the mainshaft over and insert the first-gear sleeve anti-rotation pin. Inset: later units use a ball.

Install first gear and the rear bearing. Press on the upper fifth gear as well. If you have a 0.80-inch OD fifth gear that doesn't fit through the rear bearing race, install the mainshaft assembly in the case without the fifth gear on the shaft.

Once the fifth gear is installed, always follow up with installing its snap ring. For some reason, I find people leave this step out.

Install the first-gear roller bearing and sleeve assembly and make sure you catch the rotation pin/ball in its slot.

This is how you press on a fifth-speed gear. The gear does not fit through the rear bearing race. The rear bearing race is held in place with a retainer I made. The whole transmission is fitted into the press. I'm pressing against the tip of the mainshaft. The background image shows the more common approach.

This is what a completed T5 gear train should look like. Notice the direction the synchro sliders are facing and how the 3-4 slider's tapered edge is facing towards the front.

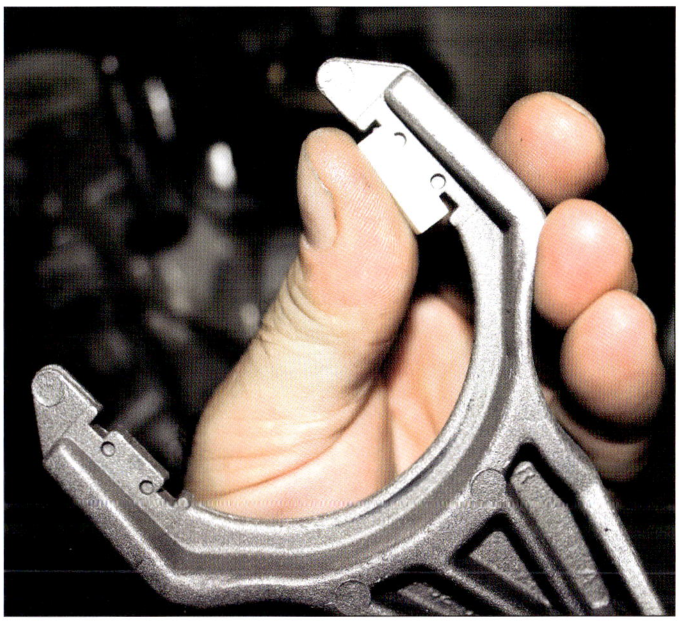

These inexpensive plastic pads on the shift forks simply snap on and off and should be replaced. This stage is a good time to replace all pads on all forks.

The fifth gear on the left shows the later-designed and stronger synchro engagement-tooth design than the gear on the right.

Drop the lower fifth gear and its new synchro ring in place. Install the fifth speed fork in the fifth gear synchro assembly, and slide the whole assembly into place. Notice the orientation of the slider, hub, and fork.

Install the fifth-gear spacer and retaining ring with the reverse brake variation install, the reverse synchro ring, brake cone, washer, and retaining ring. It is also a good time to install a new oil funnel.

Punch the reverse fork retaining roll pin in until flush with the top of the fork. Inset: The left fork is the newer style, which adds more support to the early fork on the right. New forks are under $20, so they are a worthy upgrade.

There are many T5 speedometer drive-gear variations. The more common ones just fasten with a spring clip. Others use a ball and snap ring.

If you don't have a press, gradually walk the bearing down with a punch, working the punch from side to side.

Input shaft preparation: Press on the input front tapered bearing. Load the inside needles and flat roller bearing. Use assembly lube to hold them in place. Pocket bearing units simply have to have a race pressed in.

Install the main drive-thrust washer and needle spacer. Some older units didn't have a spacer. It keeps the needles from skewing. So I install it on every unit. Pocket bearing units don't require this step or parts.

Install the new fourth-gear synchro ring. Hold it in place my putting assembly grease in its key slots and press it into the synchro assembly.

The input shaft has a clearance slot on its engagement teeth, which clears the countergear. Align the clearance space up with the countergear and install the input shaft into the maincase.

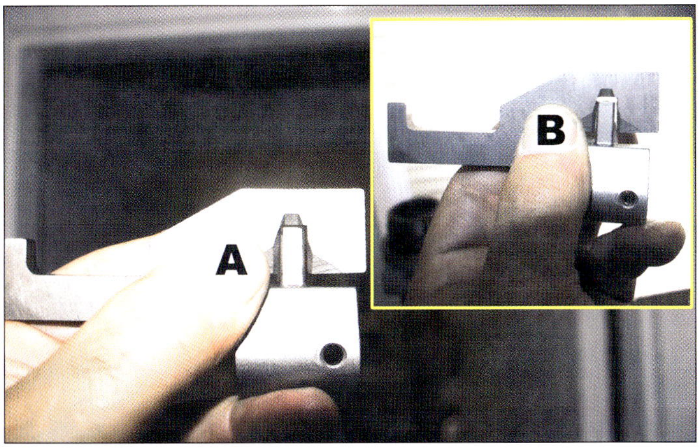

Before assembling the top cover, inspect your shift lugs. Shift lugs often stretch out. Example A shows a good lug fit, while Example B shows a poor fit. A poor fit will cause strut keys to pop and cause over shift issues.

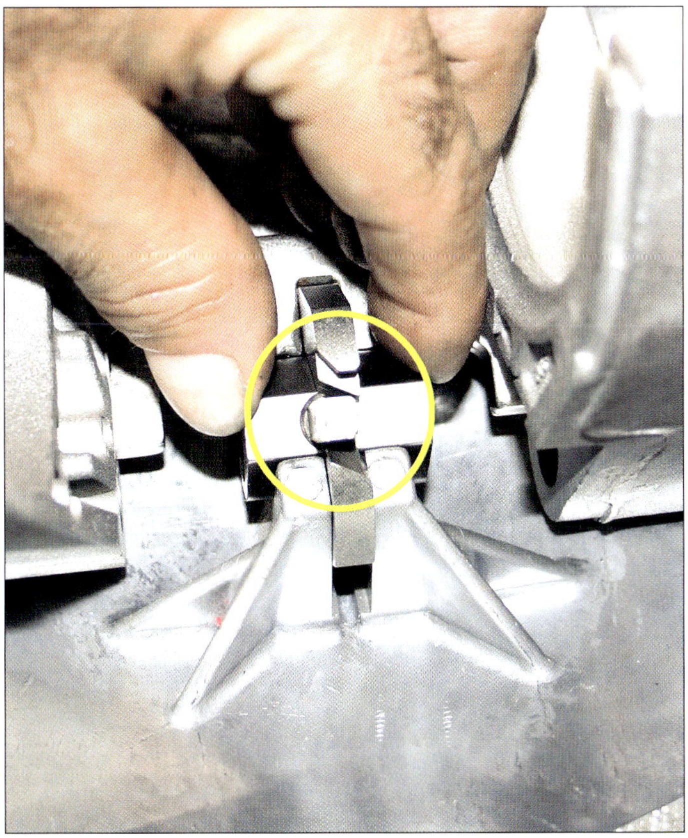

Also an important detail is how the interlock plate is located. It is very easy to install it upside down.

Since all that is required to disassemble the top cover is to punch out the shift finger roll pin and pull it all apart, most people forget how it goes back together. Notice that the 3-4 fork lug sits lower than the 1-2 fork lug. Also notice the position of the shift finger's roll pin. The front of the cover is to the left.

Apply sealant on the top of the main case, the tailhousing for the tail section, and front retainer for the retainer. This step ensures that sealant is not displaced during assembly, which could cause of a potential leak.

The fifth-gear-reverse arm rests on top of the other lugs. Install the top cover by moving it away from yourself to clear the fifth-reverse arm, and then towards yourself and down.

Use assembly grease to hold the detent ball in the tailhousing's track plate.

First install both dowel bolts, and then the remaining eight bolts, to fasten the cover. Torque the bolts to 12 ft-lbs.

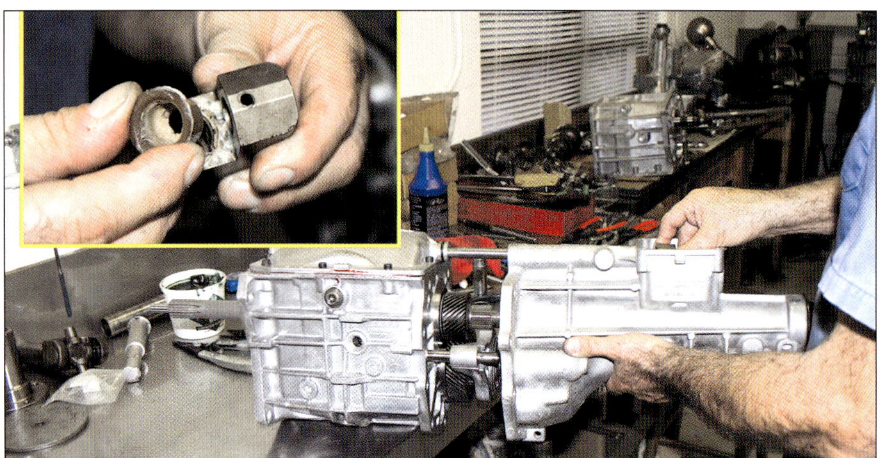

I've applied sealant to the tail, loaded the detent spring in the offset lever, and installed a new shifter bushing. Both are held in place with grease. I position the offset lever's spring on the detent ball in the tail in the neutral position. I slide the tail on while applying a firm downward pressure on the offset lever. Torque down the tail's eight bolts to 25 ft-lbs.

The last step is setting up input-shaft clearance. I shoot for zero endplay with about 0.003-inch preload max. Peelable T5 shims manufactured by 5speeds.com make the job a snap. Do not install the seals. Once the endplay is set, seat both input and output shafts with a rap of a rubber mallet, then recheck for endplay.

Once you're happy with the endplay, install the front seal, front retainer, and rear seal. Older transmissions have aluminum retainers that gall and wear fast. Upgrade to a newer cast-iron retainer. Torque four bolts to 20 ft-lbs.

Final check: 1) Tailhousing bolts are torqued and residual sealant wiped clean. 2) Offset level has roll pin and shifter bushing installed. 3) Front retainer bolts are installed and residual sealant is cleaned. 4) Reverse light switch is installed.

Continued from page 49

Jaguars required a very short transmission, so basically I shortened the T5 from 24 inches to 15 inches using a custom designed mainshaft and tailhousing. It would only see 400 hp. In 10 years, who knew that because of the JT5 size, they would be installed in everything from Ferraris, Aston Martins, Volvos, Morgans, MGBs and many other makes. The JT5 has been installed in over 800 street Jaguars and has sparked interest in overseas companies to now make similar products based on the T5 platform.

The need for better gear ratios and stronger gears led to a complete redesign, featuring dog-ring-style gears, modular countergears, beefed-up mainshafts. All contained better grade alloys with all gears riding on needle bearings.

By 2001, G-Force Transmissions came on the scene, concentrating on the street/strip market. Its T5 is a great shifting transmission. G-Force has made heavy-duty 9310 helical replacement gears, helical dog-ring gears, and straight-cut gears as well. They claim that these gear sets if set up correctly can handle over 500 hp.

By late 2005, Liberty's High Performance Products took over our Enduro design and made a gear set suited for drag racing. Unlike other aftermarket T5 gear sets, the Enduro mainshaft is much larger in diameter and eliminates the common areas of flex and stress. It is the only gear set that doesn't duplicate the poor factory mainshaft design.

The JT5 shortened design with a custom hydraulic release bearing.

JT5 Enduro gear set with modular countergear and removable dog rings.

The Enduro countergear is pictured above and the stock T5 countergear is below.

This is a modular helical G-Force countergear for a T5 transmission.

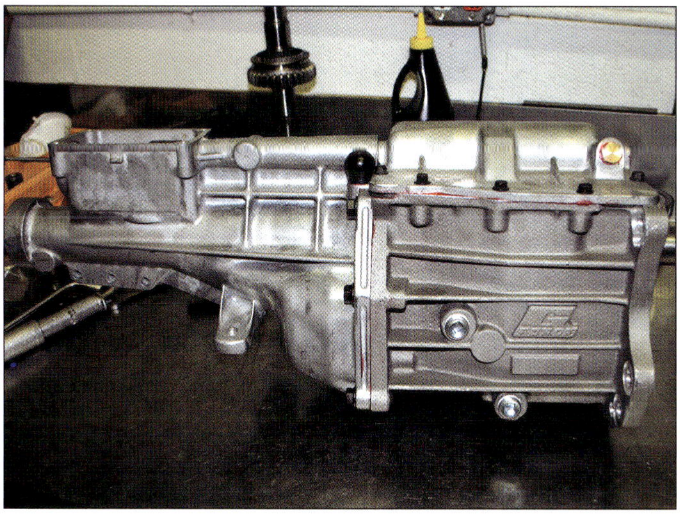

Here is a JT5 Enduro transmission built with G-Force's new heavy-duty T5 case. You can get this case in standard Ford Toploader, GM, and Ford T5 bolt patterns.

The 9310 Nickel G-Force countergears are machined from billet bar stock.

G-Force makes cast-bronze and billet-machined shift forks for the T5. This transmission has a JT5 Enduro gear set loaded in a G-Force case.

CHAPTER 4

CHRYSLER A833 4-SPEED TRANSMISSION

In 1964, Chrysler introduced the A833 4-speed, a side-loaded transmission with all gears situated in the main case. The T10 and Muncie 4-speeds use a reverse gear set outboard of the main case. The outboard design requires an additional mid plate. Use of the additional mid section can create stability and alignment issues. The A833 not only eliminated the mid section, but it added a larger center-to-center distance to the gear set. The T10 and Muncie had 3.25-inch center-to-centers. The A833 was elongated to 3.375-inch. This made the A833 the strongest 4-speed in its class. It wasn't too long before people starting adapting this transmission into other makes.

The 18-spline input is above and the 23-spline input is below.

There are many variations of this transmission. Input shafts came with 23 or 18 splines. Output shafts in early versions had 23, 26 or 30 splines. Only 18-spline inputs came with 30-spline outputs. These were typically dubbed "Hemi" boxes.

Chrysler's New Process Gear was responsible for making the A833. It was also the only transmission ever offered from the factory with slick-shifted gears. Besides incorporating the slick-shift process, race "Hemi"-style transmissions came with better alloy gears and bronze bushed speed gears to reduce frictional losses as well as gear to mainshaft seizures.

These are variations of A833 cases. Note the different extension housings and bearing bores.

Chrysler A833 Transmission Ratios

1964–1975 Chrysler A833

First	Second	Third	Fourth	Reverse	Engine Size
3.09	1.92	1.39	1.00	3.00	225/ 273/ 318
2.81	1.93	1.39	1.00	2.57	426 / 440
2.59	1.77	1.34	1.00	2.36	426 / 440
2.65	1.61	1.39	1.00	2.57	426 / 440
2.65	1.93	1.39	1.00	2.81	8 Cylinder, except above
2.44	1.77	1.34	1.00	2.40	8 Cylinder, except above

1964–1975 Chrysler A833 Race

First	Second	Third	Fourth	Reverse	Color Code / Engine
2.65	1.80	1.39	1.00	2.73	426 / 440
2.65	1.64	1.29	1.00	2.73	Green Stripe
2.14	1.43	1.19	1.00	N/A	Yellow Stripe
2.65	1.93	1.39	1.00	2.81	White Stripe
2.65	1.64	1.19	1.00	N/A	Red Stripe
2.65	1.57	1.29	1.00	N/A	Blue Stripe
2.65	1.75	1.19	1.00	N/A	Purple Stripe

1975–1980 Chrysler A833 Overdrive

First	Second	Third	Fourth	Reverse	Engine Size
3.09	1.67	1.00	.74	3.00	225 / 318
3.09	1.67	1.00	.72	3.00	225 / 318

Typical bushed race gear. Notice the paint marking, used to ID the type of gear.

Here is a Hemi countergear with purple paint. Also notice the machined grooves for ID and assembly purposes.

The Super T10 4-speed first gear on the left isn't really "super" compared to the larger Hemi first gear on the right.

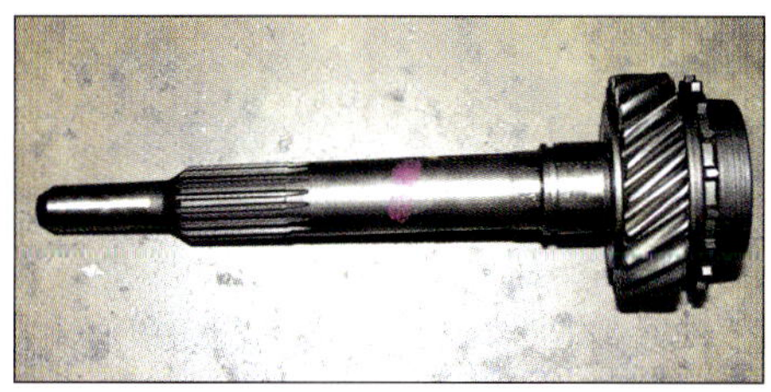

Here is a factory race Hemi input with factory slick-shifted engagement teeth.

A 1-2 slick-shift slider with a speed gear inside shows how the overall shift gap is increased.

A company called Passon Performance has reproduced just about every part of the A833 transmission, which has been a major undertaking since so many variations exist. Popular race sets are available. All case configurations have been meticulously reproduced in aluminum for weight savings because an iron A833 can tip the scales at 125 pounds.

General Motors also used aluminum and iron versions of the A833 in 1981–1985. Many parts from these will interchange with older units. Typically found in GMC Sierra pickups, the A833 was renamed the NP440 and has been a popular 4-speed conversion for GM muscle cars because it has the same bolt pattern and splines as a Muncie 4-speed.

A twisted Hemi input shaft. These are extremely heavy-duty shafts that require enormous abuse to bend. However, where there is a will, there is a way!

Here is a new Hemi 4-speed.

Passon Performance cast every tail combination in high-strength aluminum alloy.

These are original cases. Notice how the factory alloy case was beefed up to handle the remaining weight of the unit.

S-A DESIGN

GM/Muncie 4-Speed Transmissions

The Muncie 4-speed was GM's answer to the T10, and by mid-1962 full working prototypes had been built. It replaced the T10 in General Motors' cars, starting in February of 1963. Corvettes saw installation by May through June of 1963. It was based on the T10 design, and I have been told that BorgWarner engineers helped design it. In fact, physically first- and second-speed gears were identical in tooth count and pitch. Some of the later versions of the T10 had 9310 Nickel alloy gears. Because the first-speed gears were identical to the Muncie except for internal bore and synchro cone sizes, a company called Stahl made an adapter sleeve to use the stronger T10 first gear in the Muncie. The first Muncie in 1963 had a "thin fin" tailhousing and a 6207-style bearing in the front. First gear rode directly on the mainshaft. By 1964, first gear rode on a sleeved bushing and the front bearing diameter was increased, and so was the webbing on the tailhousing. As power levels climbed in 1965 and GM started putting big-block engines in cars, the countershaft diameter increased from 7/8 to 1 inch by 1966. The Muncie had several-ratio, case configurations, and spline counts. The last year GM offered the Muncie was 1974. However, some early 1975 cars built in 1974 had Muncies. The Super T10, an improved version of the T10, was reintroduced in 1975 GM cars. Power levels dropped in 1975, requiring different gearing. Since the ST10 had better ratio options, it probably was cheaper for GM to purchase complete transmissions rather than tool up to make different gears for the Muncie.

How to Identify a Muncie

People commonly call a Muncie an M21. Or they think a Muncie passenger-car transmission is the same as a Muncie truck transmission. They are not. What do M20, M21, and M22 actually mean? The "M" codes are RPO (Regular Production Option) codescodes. They don't mean Muncie. Below is a list of some standard "M" RPO codes. You can have an M21 4-speed in a 1979 Camaro, and it is actually a close-ratio Super T10.

RPO M Code Transmissions

Model No.	Type
MC1	Transmission; 3-speed
M11	Shift Lever; floor mounted
M20	Transmission; 4-speed wide-ratio
M21	Transmission; 4-speed close-ratio
M22	Transmission; 4-speed close-ratio heavy duty
M35	Transmission; Powerglide automatic with 6-cyl
M39	Transmission; 5-speed BorgWarner
M40	Transmission; Turbo Hydra-Matic automatic with 6-cyl

To correctly identify a Muncie, you need the following. First you'll need a main case casting number. Second you'll need a spline count of both input and output shafts and confirmation of any rings or grooves exist around the input shaft splines. Third, you'll need the date codes and VIN numbers that help confirm that the above case and gears belong together. Tooth counts on the input shaft help confirm a certain gear ratio. However, you may not have access to this information if the transmission is still in the car, or a vendor at a swap meet is unwilling to remove the cover. Sometimes replacement inputs have no ID grooves.

Please note that there are seven different Muncie input shafts. All 26-spline inputs came with 32-spline output shafts and all 10-spline inputs came with 27-spline output shafts. A common mistake is thinking that all "fine spline" 26-spline input shafts are M22 heavy-duty types. This is not true. An M22 gearbox has about a 20-degree helix angle on the gearset, rather than a 37-degree angle. Also M22 gear sets were a different heat-treated alloy. The straighter angle was designed to produce less end loading of the geartrain and less heat, but it created more noise, thus the nickname "Rockcrusher." The heat-treated/shotpeen process allowed for more impact of the gears. If the Muncie has a drain plug, it does not necessarily identify it as an M22. Again, the first M22 boxes had drain plugs, but all 3925661 castings had drain plugs. And by 1969, most Muncies had drain plugs.

Serial and VIN Numbers

Serial numbers for Muncie 4-speeds always begin with the letter "P." P stands for Muncie plant. Since the Muncie 3-speed used the M-designator, the letter P was used to identify 4-speed transmissions assembled in the Muncie plant. P does not mean passenger car. If P stood for passenger car, you'd see it stamped on every 4-speed, including a T10, but it is not. See the list below of GM plant codes. The serial number is the date code the transmission was built for a particular production year. Serial numbers from 1963–1966 included only the month and day. P0102 indicates that the Muncie was built on January 2. From 1967–1968, the serial number got a year designator added, and a letter designator for the month, such as P8A01, meaning January 1, 1968. One important point: a Muncie dated with a December build date was actually built the prior year. An example would be the date code P8T13. This is for a 1968 production car. The "T" stands for December, and 13 is the day. It was actually built in December of 1967 for the 1968 model year. To confirm this, simply look at the VIN number, and it will usually have a lower production sequence justifying an early build. 1969–1974 Muncies got a ratio designator added to the end of the serial number. An example would be P4D23B. This equates to April 23, 1974, M21 ratio.

Putting It All Together

You should start by looking at the main case casting number. Using the charts in the book, you can determine the date range that a particular casting was used in. Check to see if the input shaft has any rings around the splines and then check the stamped P code to help determine ratio. VIN numbers can help you confirm year, model and make as well. Later VINs have assembly plant codes and years. For example, a VIN starting with a 19N would be a 1969 Chevrolet made in the Norwood plant, while one starting with a 29N would be a 1969 Pontiac made in the Norwood plant. This could be useful for finding out what car the transmission was originally in. I always say GM never designed this system for people to accurately decipher codes 40 years later at a swap meet. You can be sure that in 40 years a great number of transmissions have been assembled out of spare parts and cannibalized. So I always like to pull the cover, look at the gears and count teeth to determine what I have.

Some input shafts produced by the aftermarket and General Motors have no identifying rings on them. The rings originally corresponded

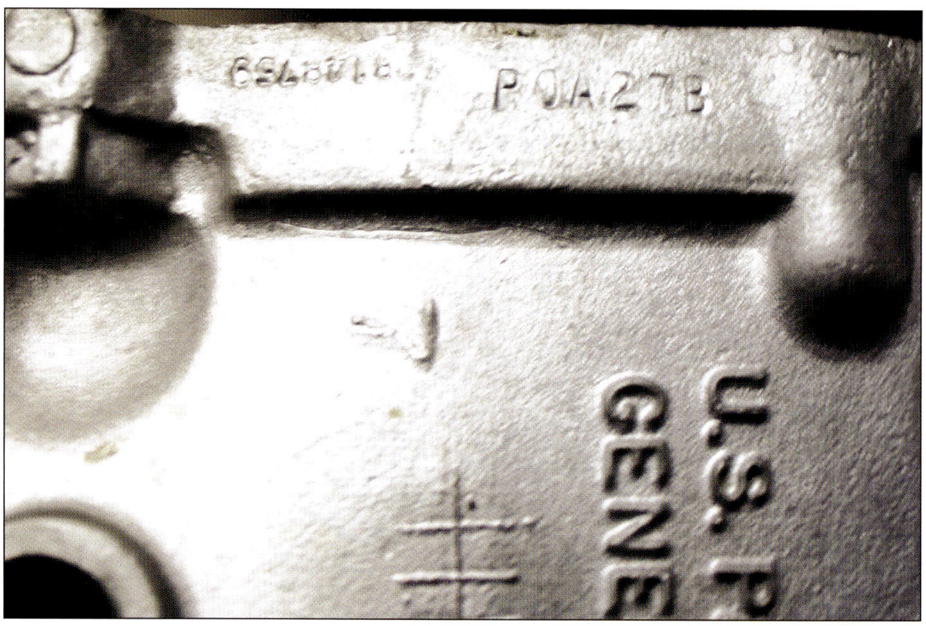

This shows the typical VIN and serial number stampings on a Muncie.

Muncie Ratio Designator

Letter Code	Type
A	M20 Ratio
B	M21 Ratio
C	M22 Ratio

Muncie Transmission Plant Codes

Letter Code	Type
S	Saginaw three-speed
C	Cleveland Powerglide
M	Muncie three-speed
T	Toledo Powerglide
H	Muncie HD three-speed
E	McKinnon Industries Powerglide
K	McKinnon three-speed
A	Cleveland Torquedrive
R	Saginaw four-speed
B	Cleveland THM 350
P	Muncie four-speed
Y	Toledo THM 350

Muncie Month Codes (1967–1974)

Letter Code	Type
A	January
B	February
C	March
D	April
E	May
H	June
K	July
M	August
P	September
R	October
S	November
T	December

with rings or grooves on the counter gear so that the assembler matched a one or two ring input with a one or two ring countergear. When manufacturing was stopped, GM stopped making inputs with these marks, probably to save machining operations.

The following images show that you cannot always use the OD grooves on input shafts as a sure fire method of identification.

The upper gear is a 10-spline M22 while the bottom gear is a 26-spline M20. Note the two grooves on the input designating the M20 ratio.

These GM gears are both 10-spline M20 inputs; however, only the bottom one has OD grooves. These are GM gears.

Muncie M20, M21 and M22 Input Shaft Tooth and Spline Count

Year	Type	Rings	Spine	Tooth Count
1963–1965	M20	0	10	24
1966–1970	M20	2	10	21
1970–1974	M20	2	26	21
1963–1970	M21	1	10	26
1970–1974	M21	1	26	26
1967–1970	M22	0	10	26
1969–1974	M22	0	26	26

Muncie M20, M21 and M22 Ratios

Year	Type	Rings	Ratio
1963–1965	M20	0	2.56 / 1.91 / 1.48 / 1.00 / 3.16
1966–1974	M20	2	2.52 / 1.88 / 1.46 / 1.00 / 3.11
1963–1974	M21	1	2.20 / 1.64 / 1.28 / 1.00 / 2.27
1967–1974	M22	0	2.20 / 1.64 / 1.28 / 1.00 / 2.27

Muncie 4-Speed Case Casting Numbers

Year	Casting ID	Description
1963	3831704	7/8-inch Bore, only small 6207NR Front Bearing, Pat. Pending
1963–1964	3839606	7/8-inch Bore, regular Bearing Pat. Pending
1964–1965	3851325	7/8-inch Bore, Patent Pending, Mostly 1964
1964–1965	3851325	7/8-inch Bore, Patent Number
1965	3864606	Patent Number
1965	3864***	Milled off last 3 digits 7/8-inch Bore Patent Pending
1966–1967	3885010	1-inch Bore, Patent Number
1968–1970	3925660	1-inch Bore, Patent Number 3925661 1970–1974 (some early 1975 cars)

Two castings rarely seen in Muncie 4-speeds are the 3839606 and the milled and unmilled version of the 3864606.

M22 Heavy-Duty Rockcrushers and the New Muncies

Alan Colvin, author of *Chevrolet by Numbers*, and Brian Higgins of SK Speed documented that 7/8-inch diameter M22 countergears did exist. Since the 7/8-inch bore was pre-1966, it suggests that the M22 transmissions were for 1965 cars. Colvin discovered a Chevrolet Engineering Change Recommendation was issued in December 19, 1964, requiring a change of the M20/M21-style transmission into the M22 because of the Grand Sport Corvette Race Program. Prototypes had the 7/8-inch countergear, but 57 M22s were built using 3851325 cases modified with 1-inch bores, added drain plugs and they had the casting number completely milled off. It's hard to imagine that 41 years later we are still designing transmissions based on the M22 concept.

I've always said the M22 was not a drag-race but a road-race transmission. Yet drag racing was much more popular in the 1960s than road racing. Today, we either have helical gears or straight-cut spur gears. Think of the M22 gears as semi-helical. The low helix angle of M22 gears produces less heat, less side loading, and stronger load distribution across the gear tooth. It more efficiently transmits power than a spur gear set, and it is not as noisy. The whine of the M22 gear set is a characteristic of that low helix angle. Four decades ago, we could run street cars with a 4.56 axle ratio, but today this is rarely done. These came with a 2.20 first-gear ratio, and this "dead" ratio just doesn't work well today. That is why newer 5-speeds are all the rage.

By 2000, almost every Muncie 4-speed that came into the shop had badly worn gears because the main case was either worn or repaired. George Sollish, president of Autogear Co., asked me about building a new Muncie 4-speed case and what improvements needed to be made. Using input from myself and several other volume rebuilders, he designed the Muncie Supercase. The case had thicker mounting ears with improved webbing, used a wider sealed front bearing, and incorporated soft plug sealing on the countershaft. This addressed ear breakage problems and leakage issues. The case casting uses state-of-the-art techniques that yield less stress risers combined with a better heat-treated aluminum. Gradually, more and more new components were added. I suggested a nodular iron mid plate, and the Superplate was created. A new side cover followed and then new tailhousings. One of the parameters was to make every part backwards compatible, so you could bolt up any older Muncie parts to the new stuff.

By May 2003, we were building complete new Muncie-style transmissions with all new components. Specialty gear ratios were created to give the Muncie a modern street approach as well as make it more competitive on the race track. You can now get an M22 with a 2.56 first gear or an extra-close-ratio M22X with a 1.17 third-gear ratio. There are also overdrive conversion gear sets. See the Muncie build up in this book—a great deal of the recently designed parts are used in the build.

GM Speedometer Primer

With so many different tire sizes and axle ratios available today, it is not unusual to be driving around with an inaccurate speedometer. Since speedometer gear sets varied from make and model and the same model could have different axle codes, it was not uncommon for the same Muncie 4-speed to have several unique part numbers based on different speedometer gear sets.

Someone may ask: "Do you know what speedometer gears I would need for my 1967 GTO with a 3.55 rear?" Most don't know their tire diameter or don't realize that changing axle ratios and tire sizes will change your speedometer reading. This is a weekly question for me, so I thought the following quick-reference may help.

The speedometer drive gear (the one that goes on the mainshaft) commonly comes with either a 7-or 8-tooth worm gear. You determine the number of worm teeth by laying the gear on its side and counting each tooth as it comes to an end. They come in two outside diameters (1.76-inch and 1.84-inch) as well as two inside diameters to coincide with 27- and 32-spline output shafts. Units built in 1969 and 1970 had plastic versions of the 8-tooth 1.84-inch diameter gear, which is held on with a clip. I always replace these with a press on steel gear. The factory only used 8-tooth 1.84-inch diameter gears on the 32-spline shafts, but you can also get 7-tooth ones from the aftermarket. There is an early 6-tooth drive gear, which is not available for the driven pencil gear, so I always recommend getting rid of these.

Here is a photo of a 3839606 casting, which is very rare.

This is a 1965 casting with the factory milled off numbers.

Calculating Speedometer Errors

You can download equations from the Internet to determine your correct speedometer reading at a given speed with a certain tire size and axle ratio. GM vehicles that use the Muncie and T10 can make use of the below equation.

Speedometer Gear Ratio (SGR) = 63360 x (Axle Ratio) / 3141.6 x (Tire Diameter)

To determine the driven gear's tooth count, you need to multiply the number of teeth on the mainshaft drive gear by SGR.

The larger-diameter steel gear uses a smaller driven "pencil" gear. The smaller steel gear on the right uses a larger-diameter driven gear. Use the wrong combination, and you'll either have no speedometer reading because the gears are not meshing or the gears are stripping!

So, by taking a vehicle with a 3.08 rear axle and 26-inch tire, your SGR ratio equals 2.389 when using the above equation. If you have an 8-tooth drive gear, your driven gear is 19 tooth.

Here's another method, which will work for any make or model, that's pretty accurate. This is particularly useful for fixing errors when changing out rear-axle ratios, yet it retains the same tire size from the factory. The above example has a car with a 3.08 rear, 19-tooth driven gear and 8-tooth drive gear. These gears form a speedometer factor. The formula for speedometer factor is:

(Speedo drive gear / Speedo driven gear) x Axle Ratio = Speedo Factor

Use the same gears and axle ratio and you will come up with an **SF** equal to **1.296:1.**

Now you have changed the axle ratio to 3.55:1. Since you know what the original SF was and your drive gear is fixed make the following calculation:

(Drive Gear x Axle Ratio) / SF = New Driven Gear Tooth Count

Using the above formula, you now have to replace the 19-tooth gear with a 22-tooth gear as per your calculation.

Muncie 4-Speed Rebuild

Most of these transmissions have hit the 40-year-old mark. The transmission chosen for this rebuild had quite a few of the common problems associated with these old transmissions. This Muncie was of the early 1964–1965 era. Because of the age and condition of this transmission, it became a candidate for some of the newer-style upgrades. Muncies really don't need to have any measurements taken. If something is worn, replace it. Parts are cheap enough. Countergear endplay should be no more than 0.035 inch. If your endplay is excessive, then you have wear on the countergear thrust surface or maincase. You can compensate for wear by making shims out of shim stock to bring the thrust washers more in spec. I still prefer new parts.

The transmission looks fine from the outside, but do not take things at face value.

A closer look reveals a stretched-out countershaft hole, which was peened to try to make the loose shaft fit tighter.

The side cover interlock support pin (between the two shifter shafts) is an early design, which can fall inside the transmission. Newer styles have an outside support flange.

Drive the tapered lock pin for the reverse shifter shaft out from the bottom upwards. If the pin is frozen, use some heat.

Shift the transmission into second gear, remove the seven sidecover bolts, and remove the side cover. Don't worry about the shift fork or shifter shafts coming out of the cover.

A close look at the input shaft and mating countergear section also known as the head set, reveals major breakage.

Remove the four bearing retainer bolts and pry off retainer. Lots of transmissions have stripped threads in this section or oversized bolts and threads.

The front bearing "gland" nut is on backwards, which is a common mistake. Also note the cracked bearing. Lack of a pilot bushing or severe misalignment of the transmission and engine crankshaft usually causes the outer bearing race to crack.

Normally, you can pry on either side of the bearing to remove it. However, this badly damaged bearing will come out in pieces.

I always chisel off the old nut since I replace them with new nuts that are in the rebuild kits. Usually after 40 years they seize in place and are difficult to save anyway.

Pull the reverse shiftershaft out to disengage the reverse shift fork from the reverse gear. That's what removing the tapered pin allows you to do. Remove the six tailhousing bolts and pry off the tailhousing.

Remove the rear reverse idler gear, thrust washer, and idler shaft. Sometimes the shaft will stick in the tail. Notice that the lower reverse gear is a little rounded off from grinding in reverse.

Remove the output shaft and upper gear set from the main case. Sometimes you'll have to pry between the main case and midplate. Don't worry about things flying apart.

Pick out as many loose parts as possible, and then remove the input shaft. Broken inputs such as this one make great clutch-alignment tools.

Remove forward or inside reverse idler and thrust washer. If you are reusing any thrust washers, make sure that the washer tangs are still able to prevent it from spinning.

Inspect the forward edge of the inside reverse idler. The groove cut into the forward edge is due to the 1-2 slider hitting it. This causes a whirring noise heard only in first gear. Worn forks and slider grooves can aggravate this common design flaw.

The front countershaft bore of the main case reveals a severe elongation of the bore. Boring it and installing a bushing is a common fix. Since this is a 7/8-inch diameter countershaft, we can just bore the case to the 1-inch size and use a later-style head set. I'll scrap this case and use a new AGE Supercase.

The newer-style M20 input is to the left. It has 21 teeth on a larger diameter, as opposed to 24 teeth on a smaller diameter. This makes for thicker and stronger gear teeth.

The M20 countergear on the right is the new design, with 25 teeth as opposed to the early 29-tooth finer pitch. It is also takes a 1-inch countershaft. This upgrade started in 1966. The three holes in the new gear's front face are for an anti-backlash plate. We remove these plates on every M20 countergear because they have a tendency to break or loosen.

Start disassembling the main gear train by removing the 3-4 synchro snap ring. Remove the 3-4 hub and third gear. Some later transmissions may require this to be pressed off.

Here's a trick way to disassemble the mainshaft without having to pull off the speedo gear. First remove the rear snap ring that holds the rear bearing to the mainshaft, and let it rest outside of the groove.

Support the assembly under second gear and press the output shaft through the hole gear set in one shot.

The second gear, the 1-2 synchro assembly, first gear and bushing, mid plate, reverse gear, and speedometer drive gear can slide off the mainshaft.

We will upgrade this transmission to later-style synchro rings and hubs. The late style on the left has thicker rings, but a thinner hub as compared to the early set on the right. Early non-shouldered rings tend to crack under heavy abuse.

Close examination of the mid plate shows damage due to a spinning countershaft. The plate has an area to lock the pin from spinning. I'll upgrade the plate to a nodular iron AGE midplate.

Spread the rear bearing retaining ring open and tap the rear bearing out of the mid plate. Most plates have snap-ring grooves that have been widened over the years. The widened grooves cause excess gear train movement. The new nodular iron plates fit more precisely and help "girdle" the main case as well.

The front bearing retainer is grooved because the nut was put on backwards. This causes an air leak, which will cause an oil leak. It's time for a new retainer.

The second and third speed gears show fairly worn clutch teeth or engagement teeth. The synchro cones have permanent grooves, which may not allow new synchro rings to bite correctly. These will be replaced with new gears.

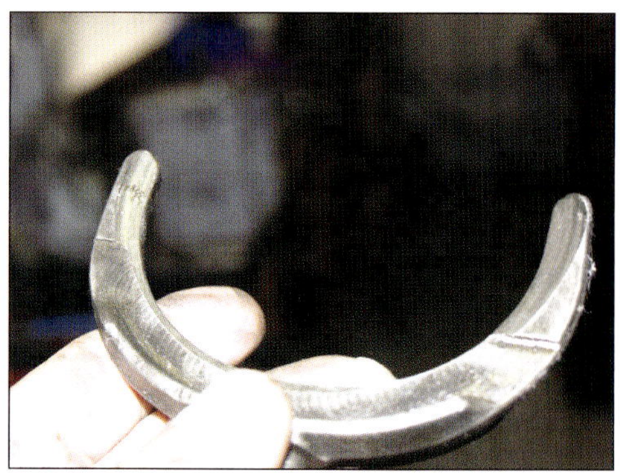

Both shift forks show wear. When forks are worn as the one pictured, shift quality can diminish. Combined with worn gears and sliders, the transmission can fall out of gear.

Here is the scrap pile from this transmission so far. Hopefully, your rebuild won't be as bad.

Remove the reverse fork, reverse shifter shaft, detent ball, and spring. Be careful not to lose the ball and spring.

Here is an easy way to remove the shift shaft seals in the side cover and tail housing.

Remove the speedometer driven gear, speedometer fitting, hold down plate, and bolt from the tailhousing.

I pry out rear seals with an old screwdriver. Most seal pullers don't work because the seals are corroded in place. You must be careful and work the seal from side to side to remove it. You can crack early tailhousings in the area because the castings are thin. Remove the bushing as outlined in the techniques section.

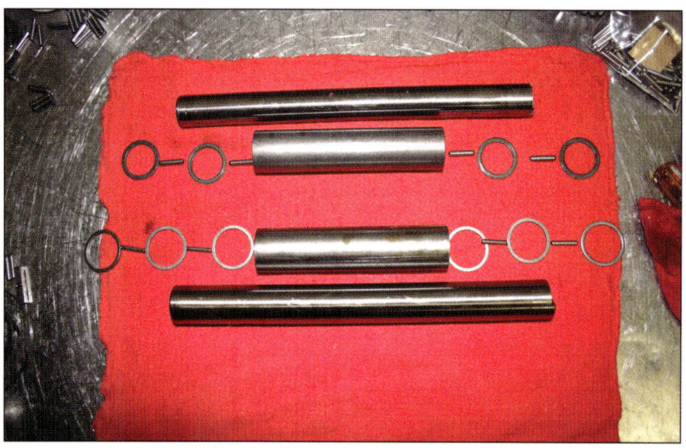

Here is how needle bearings and spacers are laid out in a Muncie. The top is the early solid spacer-tube design with four spacers for the 7/8-inch diameter shaft. The bottom is the split-tube design with six spacers used with a 1-inch countershaft. You may run into other combinations, such as a solid 1-inch tube with four spacers. The object here is to set the countergear up to have the end spacers slightly below the countergear end, usually 0.010 to 0.020 inch per side. Too much clearance can cause the needles to skew. Spacers running too tightly can jam up the needles.

The factory first-gear sleeve is to the left. It is not *worn. They all have the center tapered in for oil collection. The center sleeve is a rare M22 sleeve with milled flats to promote better lube to first gear. The end sleeve is the new AGE upgrade. The spiral shape of the oil groove forces oil into the center of the gear.*

We will install the new AGE supercase. It has thicker and stronger mounting ears. It takes a wider front-sealed bearing and has a sealing plug for the counter shaft. This addresses the common problem of Muncies leaking from the front.

Newer-style sealed front bearing is the left. This picture shows how wide it is compared to the older style.

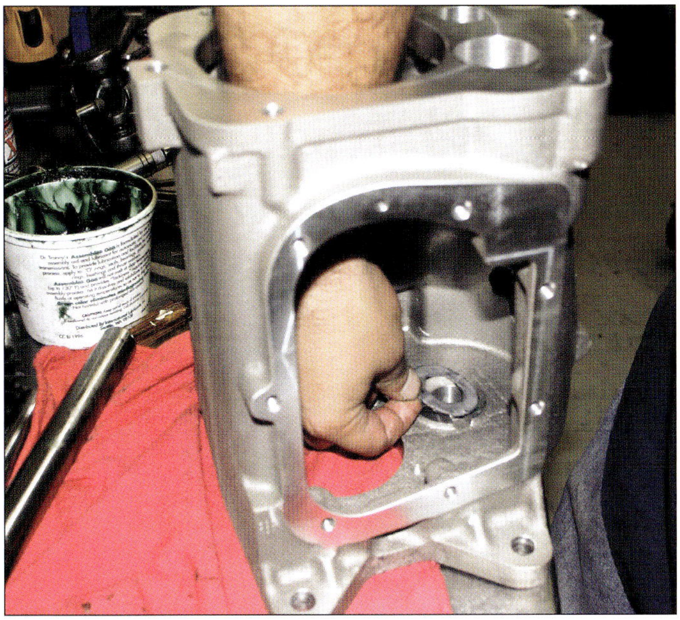

Start assembly with the case face down, and install the front thrust washer.

Drop the countergear down, resting it on the front thrust washer. Slip the rear washer in between the gear and case. This method avoids thrust washers getting pushed out of place if the countergear is sitting sideways. It also allows you to look down and see that the washers are in place. Use enough assembly grease to hold washers in place.

I install solid slingers rather than tanged slingers. Older-style slingers tend to have tangs that break off. This also acts as a 0.030-inch spacer. The new design case also corrects sloppy alignment of the input and counter shafts.

Most books direct builders to install the countergear first, but leave the countergear dropped down and install the input shaft with the bearing. This prevents you from driving the front bearing on with a hammer while the gear train is in place and damaging the fourth gear synchro ring.

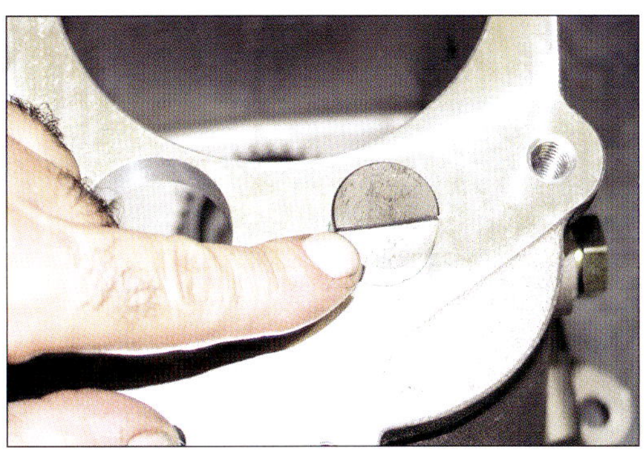

Install the countershaft with the flat side facing this position.

Here is a trick to assembling a Muncie. Place it on an old case so that you can work at a comfortable level and still keep the input shaft pointing down.

Needles must be held with grease on the input cage. They are positioned in place on the outside of the cage.

I'm grinding more of a taper to the reverse idler gear, so it never hits the 1-2 slider anymore.

The main case prep has been completed once the input, needles, and synchronizer ring, and idler washer have been installed along with a freshly ground idler gear.

Positioning synchro hubs the wrong way is a common mistake. Both hubs have their extended bosses facing towards the front.

The sliders, however, face in opposite directions. The 1-2 slider's taper faces the rear and the 3-4 sliders taper faces the front.

Install the first-gear synchro ring, first gear, the mid-plate assembly, rear bearing snap ring, and upper reverse gear.

Place second gear and its synchro ring, the 1-2 assembly and the first-gear sleeve and press it into place as a unit. Make sure that they locate correctly on the spline, and that the strut keys are indexed correctly in the synchro ring.

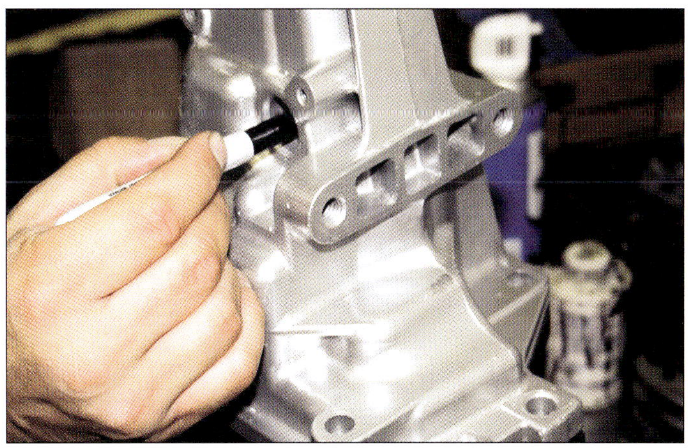

An easy way to determine correct speedometer gear position is to make sure it is close to or on center in the speedo hole of the tail. I put the tail in place and use a marker pen to mark the center point on the shaft.

The gear is heated up until it turns blue/brown and slid into correct position with a pair of pliers. Since the gears are soft, this technique avoids damage to the surface.

Install third gear, the third gear synchro ring, synchro assembly, and snap ring on the front part of the mainshaft.

Once the shaft is in place, install the shifter-shaft seal. Do this last so you're not fighting the seal's resistance against the shift shaft. This procedure actually reduces possible damage to the seal. Once the seal is over the shaft, seat it in place with a deep socket.

Position the 3-4 slider slightly forward to install the mainshaft assembly in the main case.

The tail has been prepped with a new bushing and seal. Install the reverse detent spring and ball, which are held in place with grease. Hold them down with a pry bar and slide the reverse-shifter shaft in place. You can also heat up and bend a small putty knife to make a similar tool.

Place the reverse-shift fork in place with grease, and either a nut and washer or bolt, depending on the year transmission on the shift shaft. Rotate the shaft in a forward position. The nut and bolt keep the shaft from popping out.

I'm holding the 3-4 slider in place and guiding the mainshaft over the countergear in place. This is a hard part, and you have to also make sure the fourth-gear ring indexes correctly with the 3-4 synchro assembly.

As I get over the countergear, I pull the 3-4 slider back into position, make sure the mid plate dowel is correctly aligned, then sink it all in place.

Install the rear-reverse idler, thrust washer, and idler shaft in the main case. Align the shaft so that the roll pin faces in this position. Notice my "skin" coats of sealant.

Install the tail by catching the reverse fork on the reverse gear. Once caught, tap in the shifter shaft with your palm, mate the tail to the main case, and bolt it up.

Drive the reverse shifter shaft-locking pin in from the top.

Tighten the front nut with either Channel Locks or a special front nut wrench. You can use a little thread locker on the threads. Remember, these are left-handed threaded nuts. Always use a new nut in a rebuild because this functions as a front seal. You'll have to lock the transmission in two gears at once to keep the input from spinning.

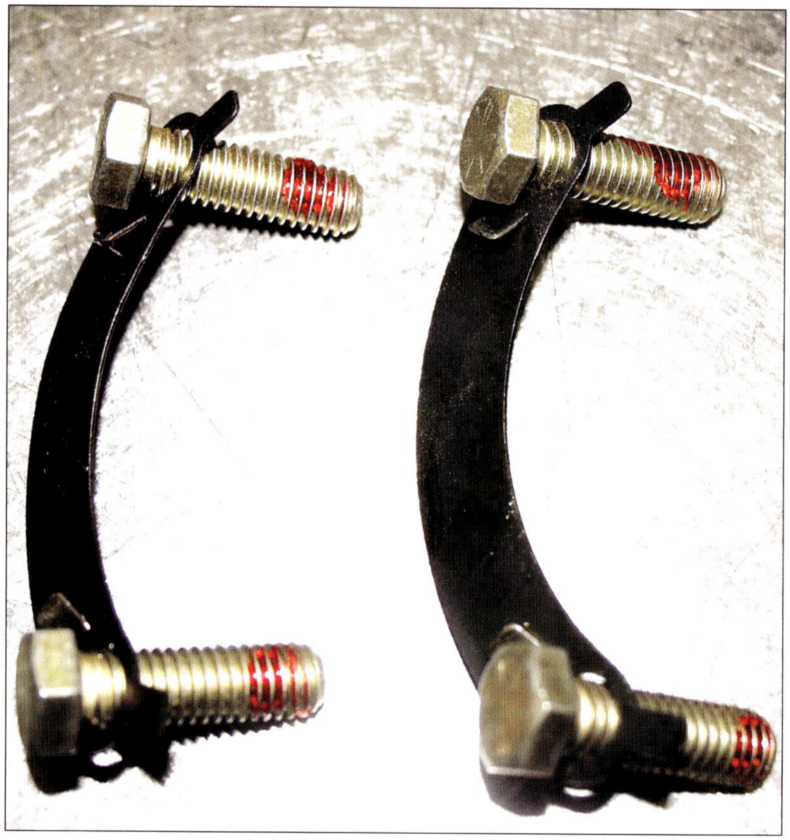

If you are using factory lock plates or stand-alone bolts, always use sealant on the threads. It acts as a thread locker as well.

The front retainer is installed with the lock plate tangs bent. The AGE Supercase uses a front soft plug on the front of the countershaft. It takes very little pressure to press it in place.

Now is a good time to stick in an old driveshaft yoke and rotate the transmission to see that everything is working smooth, not binding, and shifting. I'll always splash a little gear lube on the gears when I do this.

The new AGE side cover uses needle rollers in the side cover. Most older factory covers are extremely sloppy here. Elongated bores cause leaks and geometry problems, which lead to poor shift feel and quality. The rollers keep the shift shafts supported firmly.

I debur all detent combs on new and used units. Check for worn combs. Also note the position of the detent spring. If positioned in this manner, it can never fall into the transmission.

Place the transmission and side cover both in second gear. This allows the 1-2 fork to clear the reverse idler boss in the main case. Once in place, align the dowel in the case and tighten up the bolts.

The final step is to install the speed fitting. Older fittings had no seals. The new ones do. Don't bother fixing a rusted leaking fitting, when a new one costs under $10.

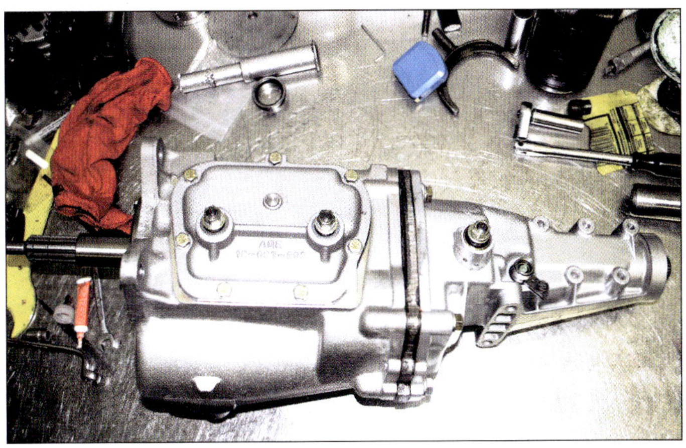

Here is the complete build. We have also used new hardware during this build. Unless I'm restoring a transmission that requires certain logos on bolts, new grade-5 or grade-8 hardware is preferable. You don't have to spend time cleaning the hardware, and you don't risk bolts breaking. Also during this rebuild please use the charts in the appendix of the book to reference torque specs.

Ford Toploader 4-speed Transmission

The Toploader's top cover and side-mounted shifter shafts distinguish it from all other American 4-speed designs. Although there were some 3-speed transmissions manufactured, the "Toploader" 4-speed loaded the geartrain from the top. The T10, Muncie, Saginaw and Chrysler A833 all used a side cover. It is also interesting to note that this transmission used a 3-shift rail design. The shift rails kept the shift forks moving in the exact same plane as the synchronizers. All other transmissions were designed with the forks moving in an arc, and these side-cover designs allowed for more shift-fork play. The Toploader design maintains a better shift-fork geometry, reducing wear and contributing to a better shift feel.

Ford, just like GM, used a T10 4-speed. The Toploader was introduced in Ford cars in 1964. There are several years in which both the T10 and Toploader were actually used within the same platform. The passenger-car Toploader came in three different case lengths and several spline configurations. Output shafts in early 1964–1965 cars had weak 25-spline output shafts. Later these were revised to 28-spline output shafts. Engines with 200-ci up to 390-ci got 1-1/16-inch diameter, 10-spline input shafts whereas the 427-, 428-, and 429-ci engines came with a huge 1 3/8-inch diameter, by 10-spline input shaft. In fact, this is the largest-diameter input shaft ever used in a passenger car manual transmission. The "big input" mainshafts also had 31-spline "big-output" transmissions. There are many variations of main cases and extension housings sporting individual and multiple bolt patterns for shifter and engine mounting. By the mid 1970s Tremec took over manufacturing this transmission for Ford. Overdrive versions appeared as early as 1977 in the Ford Granada or Mercury Monarch. Overdrive versions with shifters mounted on side turrets as well as top covers were made with alloy cases.

Identification of Toploader models can be confusing. After 40 years, few transmissions actually have remained untouched. Most units I find have been built from bits and pieces, making it harder to distinguish what was or was not original. The Toploader had over 100 different models based on engine, vehicle, and year. Two basic ratios were offered. The wide ratio features a 2.78 first, 1.93 second, 1.36 third and direct fourth. The other was the close ratio with a 2.32 first, 1.69 second, 1.29 third, and direct fourth. The close ratio was the only ratio offered with the large-style input and

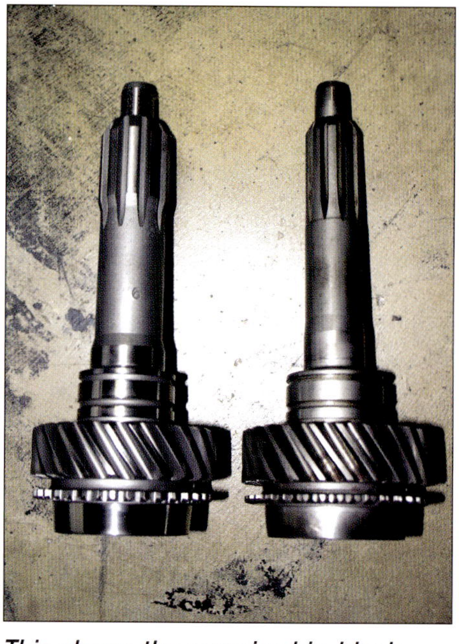

This shows the massive big-block input shaft, which can handle incredible horsepower loads.

output shafts. Today, you can purchase aftermarket wide ratio input shafts in the larger style.

Tags fastened to the passenger side of the transmission tell what make and model the transmission was used in. Tags starting with the letters HEH- followed by two additional letters are typical of early 1964–1966 Models. The RUG designator was introduced in 1967 and continues into 1986. All RUG models from 1977 and up will be overdrive models. Top cover versions used in Bronco will have the RTS designator.

Left to right, 25-, 28-, and 31-spline output shafts.

This shows a common 25-spline mainshaft failure. Often, these early 25-spline designs did not survive the shock loads of power shifts nor starting line clutch dumps. The 28- and 31-spline mainshafts were a big improvement.

Transmission Build-Up

I decided to build up a Ford Toploader using the lasted upgrades available. In this build we are going to transform a tired rusty transmission into a lightweight powerhouse.

The bearing retainer and top cover are removed. The front bearing had exploded in this transmission, and all that was left of it was the inner race. This ruined the front bearing bore of the case.

With the top cover off, some units will have the 1-2 detent spring exposed. Sometimes it is held down with a screw. Remove the spring and fish out the detent with a magnet.

Remove both set screws for the 1-2 and 3-4 forks.

Remove the four tailhousing bolts and slide the tailhousing off. Be careful not to damage the plastic speedo drive gear.

Remove the rear bearing main shaft snap ring.

Remove the 3-4 detent bolt, spring, and detent. Again fish them out with a magnet.

Remove the speedometer gear-snap ring, and slide the gear off. Use a magnet and remove the anti-rotation check ball that is under the gear.

Once the set screw and detent have been removed from the 1-2 shift rail, slide the rail out of the back of the case.

Flip the transmission over so it is lying top down. This takes the load off of the countergear and allows the shaft to be easily removed. Don't be concerned about needle bearings falling out of the case. New ones are in the rebuild kits. If you are reusing them, you may want to lay the whole assembly in a flat pan.

Punch the countershaft out of the case from front to rear.

Flip the transmission back over. The countergear will drop down, allowing the input shaft to be pulled out of the front. Tilting the whole transmission front face downward may help keep the main drive needle bearings from falling out of the input shaft.

I use a soft hammer to push the mainshaft out of the rear bearing. With the countergear dropped, you can almost get the rear bearing completely removed.

Use a bearing clamp to support the rear bearing. This will allow you to push the mainshaft out of the rear bearing completely.

Pry the rear bearing out of the case using screwdrivers or pry bars.

Once the rear bearing and front input are out of the case you will gain the necessary clearance to remove the 1-2 shift fork.

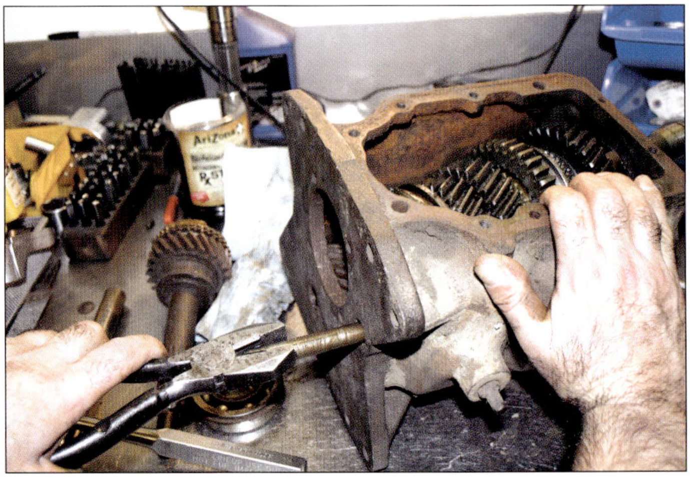

Remove the 3-4 shift rail by tapping it from the rear and pulling it out from the front. Once the rail is removed, the 3-4 fork can be removed as well.

The 3-4 rail has an interlock pin. Make sure you don't lose this pin. The interlocks that sit above and below this rail may fall out as well. If you don't want to remove them at this time, stuff the center rail support with a paper towel.

With both forks and rails removed, it is now possible to remove the entire gear train. Be careful not to catch your fingers in the case, since these parts are very heavy.

Now that the gear train is removed, the countergear can be removed as well.

Unless the reverse idler gear is damaged, rarely does this assembly need to be removed during a rebuild. To start, remove the rail set screw.

Spraying the detent/interlock bores down with WD40 may allow you to remove the detents easily. Oil deposits and sludge usually can make them difficult to remove.

Remove the reverse rail by rotating it 90-degrees clockwise to dislodge it from its own detent. Pull the rail and fork out of the case. Remove the reverse detent and spring.

Since all the rails are removed, it is now possible to remove all the shifter shafts. Please mark which ones go in the correct places.

Remove the reverse idler rail by tapping it from the front and then by pulling it out from the back. Notice the orientation of the sliding idler gear in relationship to the main idler when removing this last component from the case.

Most building errors are caused by putting parts back in the wrong place. Notice the difference between the 3-4 fork on the left and the 1-2 fork on the right.

Pull the first gear and first gear thrust washer off of the mainshaft. Notice the direction of the 1-2 slider.

Remove the 3-4 synchro snap ring, 3-4 synchronizer, and third gear. Also inspect the tip of the mainshaft for galling and wear.

Remove the second-gear snap ring, snap-ring spacer, and second gear.

Remove the 1-2 synchro snap ring and 1-2 synchro assembly. Notice the orientation of the 1-2 synchro hub. The thin edge faces towards second gear.

The most common mistake in Toploader rebuilds is mixing up the synchro strut keys or dogs. The larger key on the right is the 1-2 key. The 3-4 key is on the left. This is hard to see. New keys are notched on the 1-2 to distinguish them from the 3-4 key. Putting the long key on the 3-4 synchro will cause the unit to lock up on final assembly.

Another overlooked wear point is the key slots. The keys over time pound out of the slots. The pounded-out key allows the synchro ring to rotate farther than it should, and this causes a block-out shift condition. The unit will require a completely new assembly on both 3-4 and 1-2 because of this.

Both second and third gear have badly worn and broken engagement teeth. This is very common in the Toploader design. The new modern gears have longer teeth with torque locking angles.

The input shaft front bearing has badly damaged the gear teeth, and this gear is just too rusty for my taste.

Final Assembly

Assembling the Toploader is basically a reversal of our disassembly techniques. There are a few different detent pins and interlock pins. I have noticed that some units have different shaped detents with a flat edge compared to conical shaped interlock pins. Other units have the same size conical shaped interlocks and detents and others may be of have different lengths. It is best to lay everything out on a flat surface and pre-fit the rail system to make sure it works properly. I find it is better to test the rail system out first without anything in the way so that it is quite easy to make corrections rather than have to remove gears and start over.

Deburr and polish the shift rails. Most interlocks and detents will groove and burr up the shafts, so I debur and polish each rail and make sure they glide smoothly in the case.

We are using a new David Kee case. I have prefitted all polished and deburred shift rails in this case.

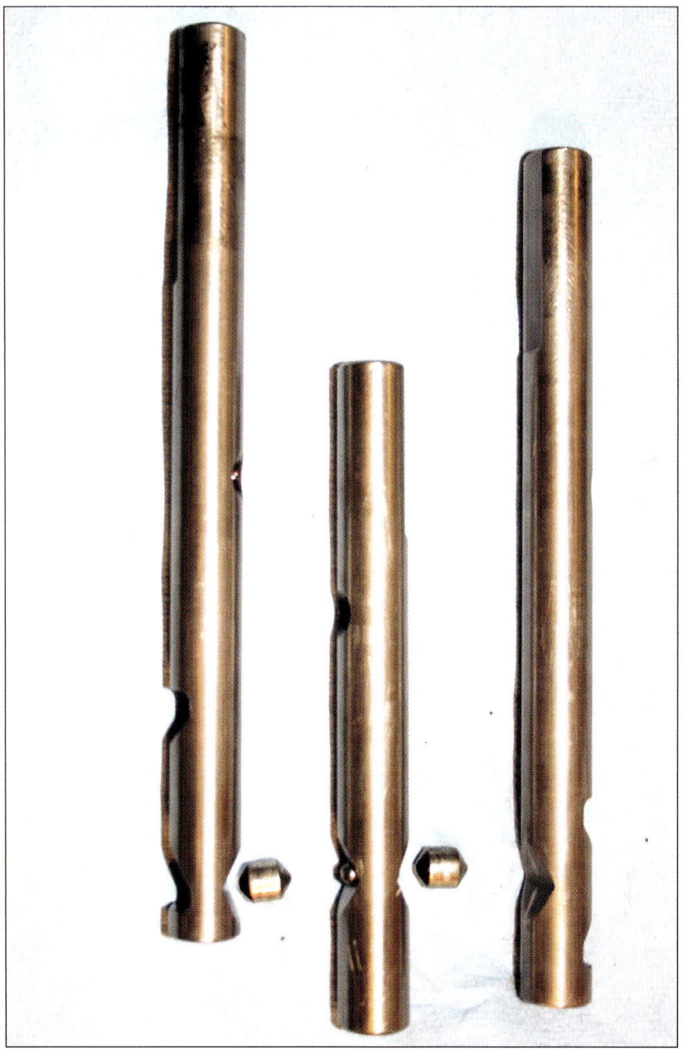

This is basically how the interlock system works. When any one rail is moved, the interlock pin is raised, preventing any other rail from moving.

David Kee makes heavy-duty shifter shafts. Note that the lower shaft is thicker than the upper (older) shaft. I've installed new O-rings on them, and they get installed in the new case first.

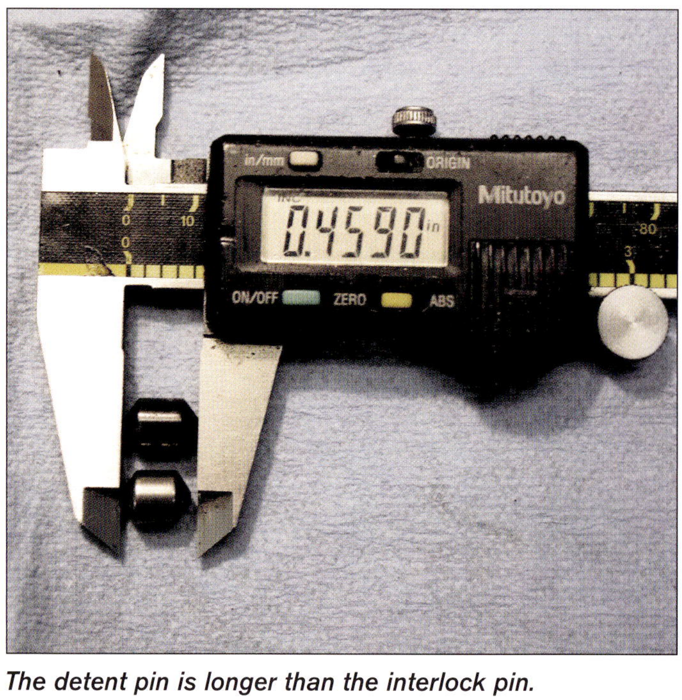

The detent pin is longer than the interlock pin. If you mix them up, the unit will not shift.

Some cases have a 1-2 detent spring that is held in place by the cover. I drill and tap the case and make my own detent hold down. I like this method better because you can fully test the transmission with the top cover off.

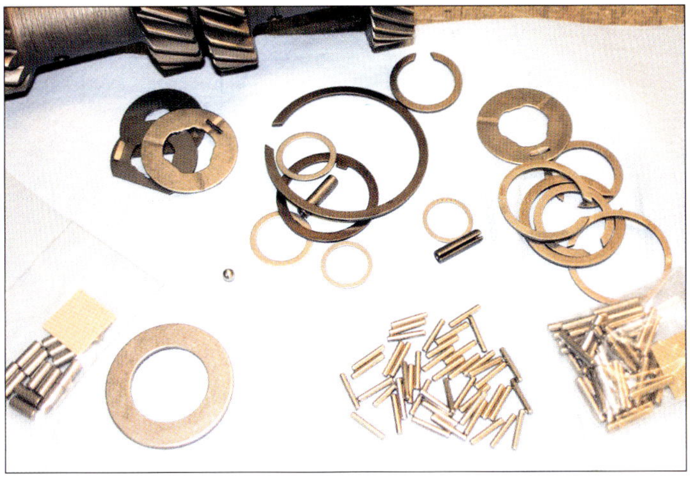

This is a typical small-parts kit sold by many companies. Never throw out your old small parts. This kit had no front bearing snap ring and the countergear needle spacers were the wrong size. The new thrust washers are nylon instead of bronze. This is actually what the newer units used, so they are fine to use.

I use this long screwdriver with a flattened tip as a tool to install detents and interlocks.

After installing the reverse idler assembly, you can install the three shifter shafts with their new O-rings. Notice that the longer reverse shifter shaft faces down.

Insert the new needles and spacers in the reverse idler assembly. Place the rear thrust washer in the case first and keep it in place with grease. It is easy to place the forward thrust washer in last after the idler shaft is supporting the idler assembly and rear washer.

I install the reverse spring and move the detent into place with my "tool."

From the top, now compress the detent down and install the reverse fork and rail into place. Once in place, reinstall the reverse fork set screw.

Before installing any other gears, it is easier to install the 3-4/reverse interlock pin. Use your special tool to get it into the bore.

I've installed the needles and spacers in the countergear and have both thrust washers installed as well. Drop the countergear in the case, making sure the washers stay aligned properly with the countershaft bore while the gear is sitting as low as possible.

Both strut springs' "humps" need to be located on the same strut key in every synchro assembly.

Now install third gear, third-gear synchro ring, the 3-4 synchro assembly, and the 3-4 synchro snap ring. Some transmissions have tighter fits than others when it comes to installing synchro assemblies.

Install the 1-2 synchro assembly and snap ring, second gear, second gear snap ring spacer, and second gear snap ring on the main shaft.

Flip the whole mainshaft assembly over and install the first-gear snap ring, first gear, and the first-gear thrust washer.

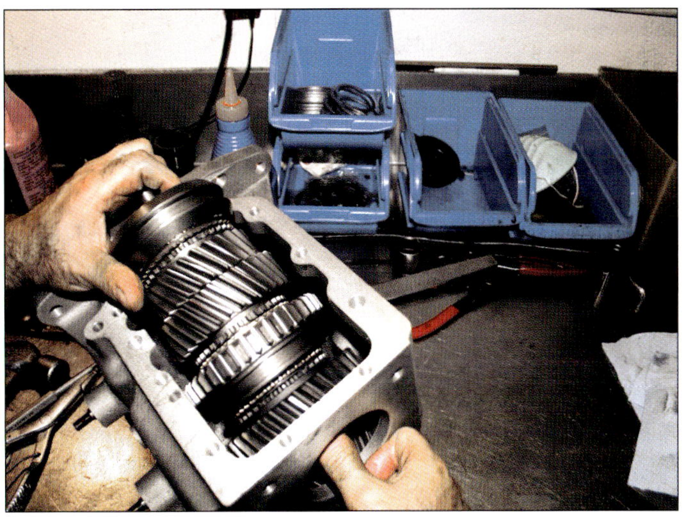

The new case as well as old cases have a relief for the 1-2 slider to clear. My thumb is keeping first gear in place.

Hold the 3-4 rail interlock in place with grease.

Locate the 3-4 fork on the shifter shaft and slide the 3-4 rail into place. Make sure to keep the rail positioned with the set-screw hole facing up.

Flip the case on its front. Use something to keep the upper gear train in place. I used an older pinion gear.

Drop the interlock in that sits between the 3-4 rail and the 1-2 rail. Position the 1-2 fork on its shifter shaft, and slide the 1-2 rail in from the back. With both forks in place, lock them in place with the set screws. Factory screws had nylon locking inserts, which may be worn. Use blue liquid threadlocker instead.

You can use a punch or pipe to press the rear bearing in place. Install both inner and outer snap rings.

Autogear has reproduced the common seven-tooth black nylon gear in steel. The plastic gear strips out all the time. It saves you the headache of pulling a transmission for a $25 part.

Install the check ball on the mainshaft. Slide the speedometer gear in place and secure it with a snap ring.

Install the fourth gear synchro ring from the front. I usually apply assembly lube on the ring and stick it in place.

The input shaft has some clutch teeth removed, so it can be installed from the front and clear the countergear. Position it so that the flat spot is facing the countergear.

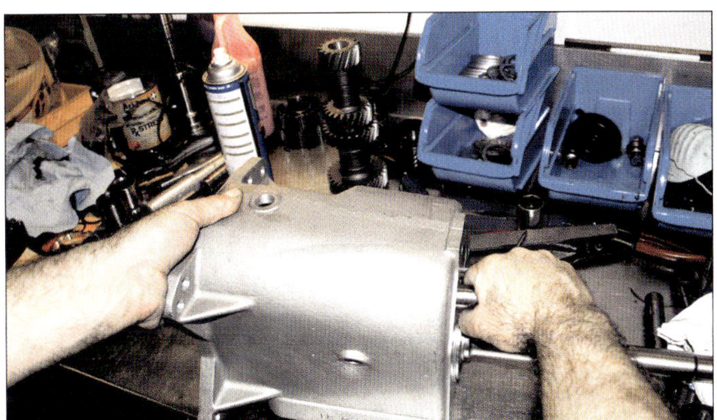

Lay the transmission on its cover side. It makes installing the countershaft simple and fast. I'm using my left index finder to hold and feel that the countergear is in place.

My finger points to the 3-4 detent, and the screwdriver is on the 1-2 detent. Use sealant on the 3-4 detent bolt.

Some Toploaders require sealant on the extension housing bolts. Torque these bolts to 35 ft-lbs.

The factory front retainer bolts have a serrated locking thread. These ruin the case threads, so regular grade-8 bolts were used with blue liquid threadlocker. Torque these bolts to 18 ft-lbs. Don't forget to install the 11/16-inch case plug.

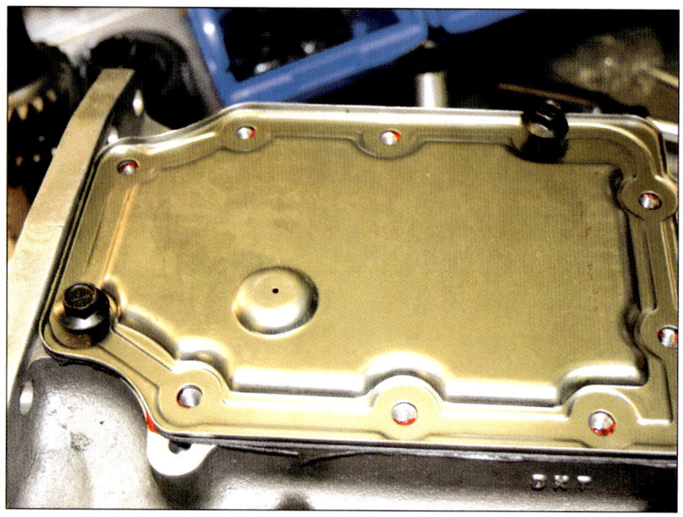

The topcover has bolts of two different lengths. The two longer bolts go in these positions.

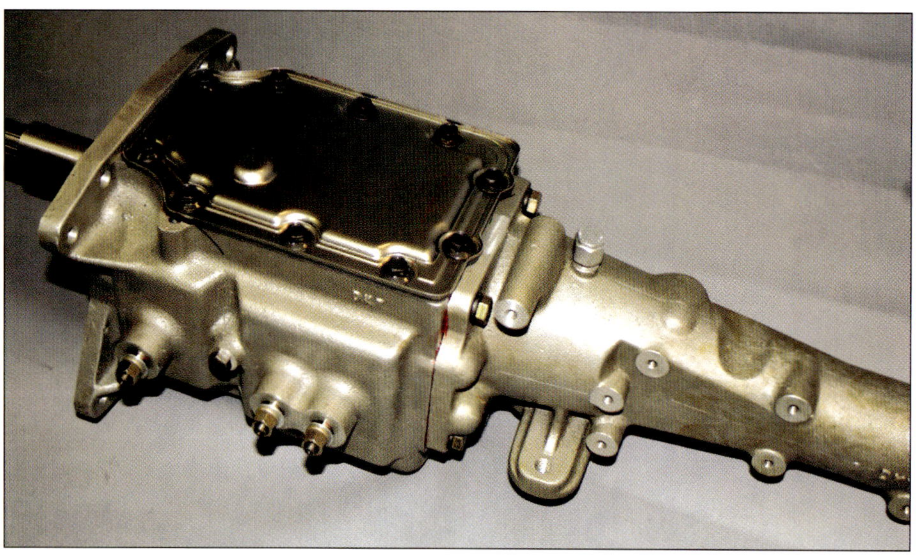

Here is the completed unit. Notice the special filtered vent I installed in the tail. The factory vent is that little bulged pinhole in the topcover.

OTHER COMMON MANUAL TRANSMISSIONS

Although there are some transmissions that may or may not be popular or suited for high-performance use, they need to be mentioned. There are also newer transmissions that are replacing the older muscle car ones that deserve a spot here as well.

A Muncie mainshaft is pictured above the weaker Saginaw 4-speed mainshaft.

Saginaw 4-Speed Transmissions

This is a GM light duty car transmission used from 1966 to 1985. It has a cast-iron case with all shift levers on the side cover. Often mistaken for a Muncie, it is sometimes labeled as one on Internet sites, such as eBay.

The cast-iron Saginaw has a 7-bolt side cover just like a Muncie, but it lacks a center support or mid plate.

A Saginaw gear set is shown on the top and a Muncie on bottom. Notice how the 1-2 slider also is used as a reverse gear.

Richmond Gear 4+1 5-Speed Transmission

Originally designed and sold by Doug Nash, the fifth gear was not an overdrive, but a direct fifth. It allowed a person to have a fairly high rear-end ratio of 3.08 and still have a close ratio transmission capable of accelerating like a Muncie M21 with 4.11 gears. Richmond gear has improved on the design, making a 6-speed called a ROD as well as newer 5-speed with an overdriven fifth gear.

ZF S6-40 Transmission

ZF Friedrichshafen AG is a leading worldwide supplier of manual transmissions. Founded in 1915, this company primarily produced transmissions for both the European industrial and automotive markets. By 1988 it developed a 6-speed transmission that GM used in the C4 Corvette platform starting in 1989. It replaced the weak Doug Nash 4+3 transmission used in the Corvette from 1984 to 1988. Many people use these in other applications but tend to forget that service parts are extremely expensive.

The new Richmond Overdrive was released in 2007. Reverse is moved to the rear, and the gears are much wider than the original 4+1 transmission.

Tremec Transmissions

Tremec had purchased Borg-Warner's manual transmission division. This gave Tremec the right to produce the T5, and T56 6-speed. They were already making the Ford Toploader and various overdrive and 2-wheel-drive versions of it for Mexican and South American truck markets. A transmission called a 3550 started out as a replacement for the 170FS (Forward Shift) Toploader. It was a heavy-duty 5-speed version for that truck market. The engineers at Tremec thought that with a tailhousing change they could make it fit as a replacement for the weak T5 in Mustangs. The first 3350 used three different shifter locations and similar ratios to the Mustang production T5. However, it really wasn't any stronger, because the truck design used individual countergear speed gears held on the countershaft with a keyway. Liberty's H.P. Products started cutting extra keyways to beef these

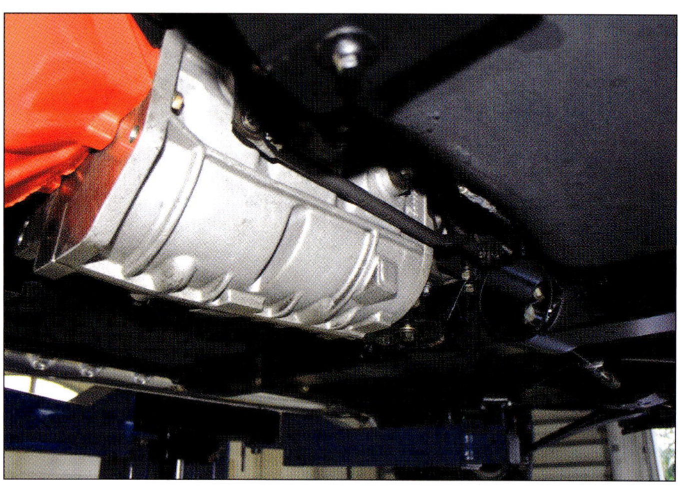

A Richmond Gear 4+1 5-speed is installed in a 1967. Note how the 3-1/2 inch center-to-center makes the transmission overhang the bellhousing.

A cutaway of the S6-40 Corvette 6-speed shows its massive gears.

Here is a nice cutaway of a TKO600. Unlike the T5 it has multiple shift rails.

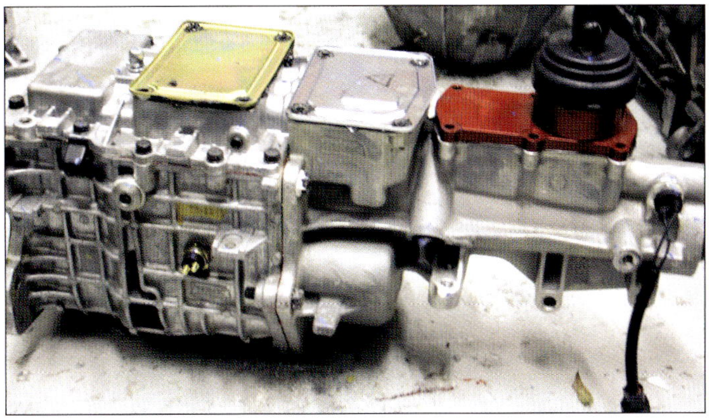

The covers are places for additional shifters. The rear shifter can be flipped around for more versatility.

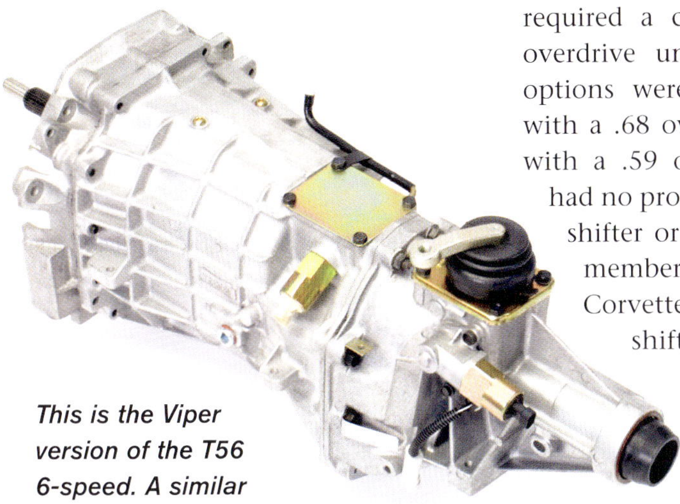

This is the Viper version of the T56 6-speed. A similar version is also in the Corvette and Aston Martins.

units up. It took several attempts to get it right. Tremec came out with a TKO version, then a TKO 500 and TKO 600. These were offered for both GM and Ford cars with mechanical and electronic speedometer pick ups and a redesigned shifter that can be flipped 180 degrees to fit both makes.

Doug Nash 4+3 Transmission

The Doug Nash Engineering 4+3 manual 4-speed with automatic overdrive was used in 1984 to 1988 Corvettes only. It was a Super T10 4-speed bolted to a 2-speed automatic overdrive. These transmissions required a computer to kick the overdrive unit on and off. Two options were available, the MK2 with a .68 overdrive and the MH5 with a .59 overdrive. These units had no provisions for mounting a shifter or a conventional cross-member mount because the Corvette used a floor-mounted shifter and a beam that rigidly mounted the transmission to the rear axle. Because the unit requires an oil cooler and a computer control, most people considering these for a non-Corvette applications should use an alternative. It is hard to find service parts as well as qualified people to work on them.

Autogear M33 "Syracuse 5-Speed" Transmission

Autogear was founded in 1945 by Irving H. Sollish and is currently under the leadership of his grandson, George Sollish. George pioneered the heavy-duty replacement cases, optional gear ratios and the manufacture of a completely new replacement Muncie 4-speed, also known as the Model 18. The M33 5-speed overcomes some unique design challenges. Perhaps the biggest one is to replace a Muncie 4-speed and keep it as close to the same footprint. Most replacement 5-speeds suffer from lackluster shifting due to increased mass combined with increased center-to-center distances. The M33 keeps the original GM Muncie centers and incorporates synchronizers on the countergear to decrease shift effort.

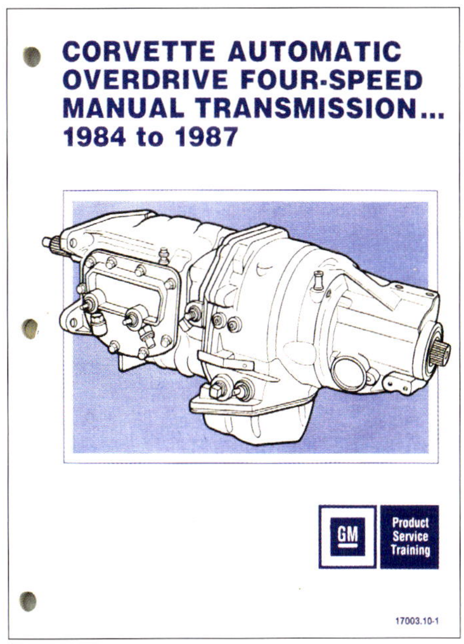

CORVETTE AUTOMATIC OVERDRIVE FOUR-SPEED MANUAL TRANSMISSION... 1984 to 1987

GM Product Service Training

17003.10-1

The 4+3 overdrive was so complex that GM published a special service manual with over 200 pages of information.

CHAPTER 8

High-Performance Transmissions

Automatic transmissions obviously shift for you, and as a result, there are no missed shifts with an automatic. If you use an automatic transmission coupled to a torque converter, you will in most cases see more horsepower loss at the rear wheels compared to a manual transmission. However, there are many factors that determine winning a race other than rear-wheel horsepower loss. Missing a shift in a drag race will cost you the race no matter how efficient your gearbox is. Unless you can step up to a dependable and consistent manual transmission for drag racing, sometimes it is cheaper and cost-effective to go with an automatic. Road racing is another story. There are many gear ratios needed to negotiate various race tracks. Daytona may require a different set of ratios as compared to Sebring. In a road-race environment, automatic design does not allow for quick ratio changes, quick tear-down, and the availability of efficient 4-, 5-, 6-, or even 7-speeds. Endurance events, such as NASCAR-style racing, require designs to be as light and efficient as possible. With the newer style of metal alloys available today combined with the state-of-the-art CNC machining and 3D modeling, gear design has evolved to a level that allows for stronger horsepower capacity within a smaller configuration.

In a typical synchronized 4-speed, such as a Muncie, T10, Chrysler or Ford Toploader, the transmission has 36 engagement or clutch teeth on the gear. The matching slider that engages the gear has the same number of teeth. When spinning above 6,000 rpm, the surface speed of the gear and slider can be very different in certain ratios. I always associate this with trying to place a stick through a picket fence while driving by that fence at 60 mph. If I'm standing still, I can stick it between the pickets, but if I'm driving by at 60 mph, the stick may bounce off the fence. However, if I remove every 10 pickets out of the fence I can get the stick through the fence at 60 mph just as easily as if I were standing still. Getting back to those 36 engagement teeth, if I remove every other tooth on the gear and slider, I've effectively cut my surface speed in half. At 6,000 rpm I now have the surface speed of a 3,000-rpm shift. The downside is that the engagement teeth must carry double the load they were originally designed for. This technique is called "slick shifting." From the 1960-to the early 1980s, it was possible to find new sliders specifically made for slick shifting. You purchased the sliders and modified your own gears. In fact, you could purchase from Chrysler complete new gear sets and sliders for the A833.

Chrysler A833 18-spline slick shift slider. Notice where the original splines were.

In the mid 1960s a tool and die maker for Cadillac was trying to improve on the slick-shifting concept. His design revolutionized the way transmissions shifted, and in fact his name became synonymous with drag-race transmissions. Joe Liberty slick shifted his friend's transmissions for drag racing. Reducing the number of clutch teeth from 36 to 18 may increase your shift speed, but it promoted more wear. So he invented the "Pro-Shift" ring. The ring decreased the number of clutch teeth to 12 but increased the width, giving them back their holding power. Business grew so fast that he didn't have the time to run the side business and work his job, so he contracted a small auto shop to do the work for him. By 1972, business was so good that Joe left Cadillac Motors and started Liberty's High Performance Products. Liberty patented the concept of the pro shift ring as a modification to a factory gear by grinding off the old engagement teeth and welding a new pro shift ring. Next he took a Chrysler A833 and modified it with clutchless gears and won best-engineered product at the 1974 SEMA Show. These transmissions were made with completely new gears, which was also a first because few people back then had the resources to do such a thing. However, the transmission was not allowed in its intended racing class because the rulemakers thought it would give Chrysler an unfair advantage.

The modified A833 was only allowed in a class that it really wasn't designed for. This led to gear breakage. One day Joe was sitting at his desk rolling around some gears and watched three gears roll against one another. He came up with a new idea, paving way for the first dual cluster gear transmission. By putting the mainshaft between two countergears, mainshaft deflection is avoided and the load of the transmission is split between the two gears. This design increased the strength by 2.8 times.

The significance of Joe Liberty's concepts and designs sometimes went unnoticed. He was also the first to "combo" gears. Grinding off the countergear's headgear is the first step of the process. To create a different ratio he cut the head gear off from another countergear and combined the two by welding them together. I sold many gears made like this and can vouch that the welds never broke. When I asked Joe how he found the talent to weld gears, he replied, "In Detroit there was an abundance of talent. You could find anyone to make anything. That's just how things were done."

Because of advanced alloys today combined with CNC machinery, a multitude of performance transmissions are made with some unique designs. Since most of these transmissions are not mass-produced every gear start as a piece of bar stock.

A JT5 Enduro gear has this removable dog ring faceplate. Notice how large the window is between the lugs. The larger window allows for higher RPM shifts.

A complete pro-shifted Muncie 4-speed utilizing new gears.

Stock Muncie input on the left and one modified with a Liberty's Pro Shift Ring.

The latest version of a dual-countergear transmission increases load capacity by putting more gear teeth in contact with one another and splitting the load in half.

Because you are using two countergears, they must be timed. I'm pointing to the timing mark. The opposite gear has to be on the exact same tooth, or the transmission will lock up.

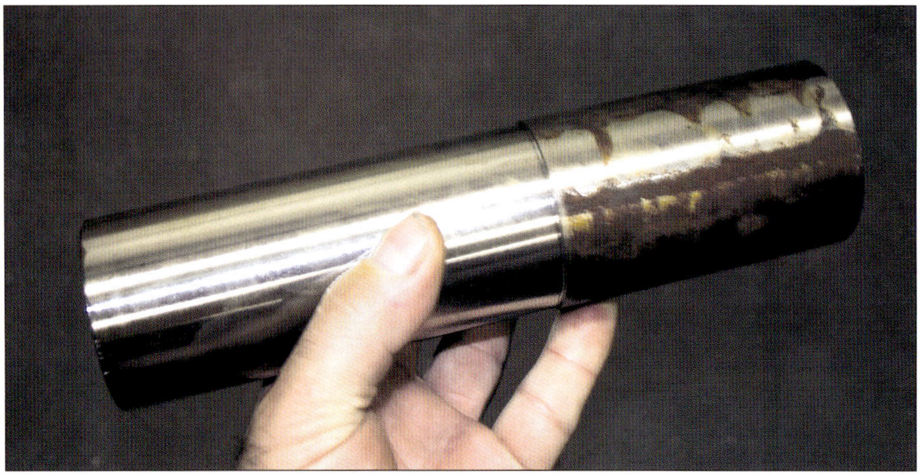

Bar stock cut for an input shaft made out of 9310VAR alloy.

A Muncie 4-speed countergear that had broken first gear is recycled by grinding the broken teeth away to make a 3-speed dirt track transmission.

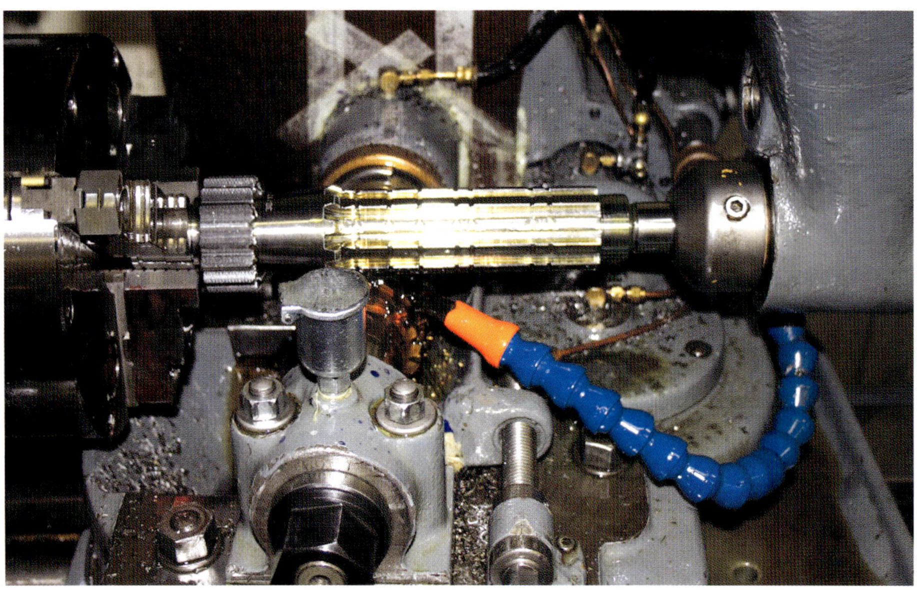

A G-Force racing countergear is being machined for a T5.

This shows the intricate machining of a Formula 1 titanium shift barrel and dog rings.

The Lenco transmission shifts by applying bands to lock the planetary gears in place. It's like manually shifting the bands of an automatic transmission.

A transmission by Quaife uses a sequential barrel with dog-ring gears.

The engagement teeth of a clutchless drag race transmission. Note how the dog teeth are ramped to pull themselves into gear.

In this Jerico air shifted racing 5-speed, air is shuttled from one solenoid to another.

A JT5 Enduro 5-speed gear set is retrofitted into a stock T5 case.

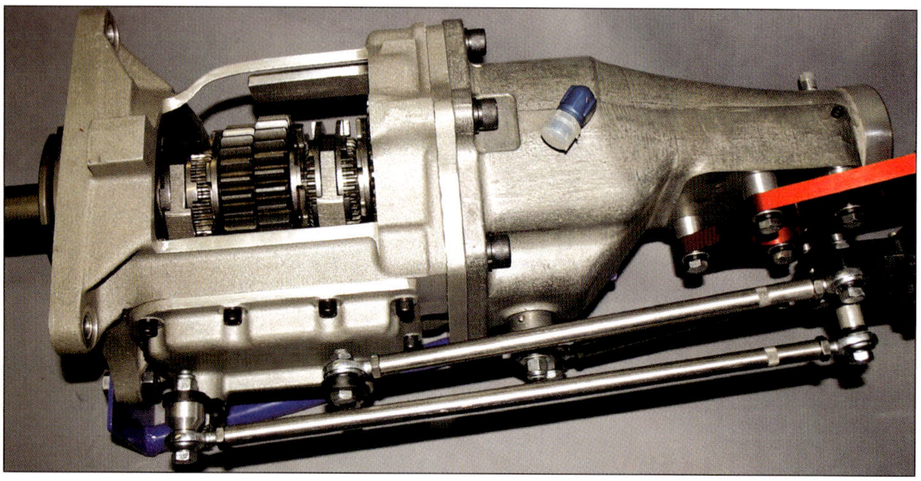

This Roltek T10 4-Speed is used for road racing with a Long Shifter.

A G-Force GF-2000 5-speed transmission. Notice the dual forks and slider for the clutchless operation. As one slider engages, the other is pulled from the opposite side.

Racing transmissions can be shifted in a standard H-style pattern or a "sequential" pattern. A sequential pattern means you pull the stick in one direction for upshifts and another for downshifts. There is no side-to-side movement; thus you don't have to worry about missing shifts because there are no gates. The concept basically is the same as a motorcycle transmission. A shift barrel rotates, and the rotation is translated into a linear movement of the shift fork.

Besides making transmissions with bigger components and stronger alloys, you can also use some of the techniques the big boys use on your everyday 4-speed. Deburr all gears to get rid of high spots that can cause the slightest gear misalignment. Rarely do you find new gears that are not dinged. In the past, people have shot peened gears to add strength. Newer processes, such as Mikronite and Rem, chemically polish and peen gears. There are dry-film lubricant coatings available and the cryogenic treatment of gears, which at times can increase gear strength like a normal heat treatment would.

The pictures shown indicate the direction of transmission design and how complex racing manual transmissions have actually become. Rarely does one get to see such a wide variety of transmissions in a single glance. At the time of this writing, very small companies have completely reproduced the Muncie, Toploader, and A833 4-speed. The costs to accomplish this would have been prohibitive 10 years ago. The quality and accuracy of the machine work far exceeds what was done 40 years ago. I expect this trend to continue as global markets open up and as unique and important designs materialize faster.

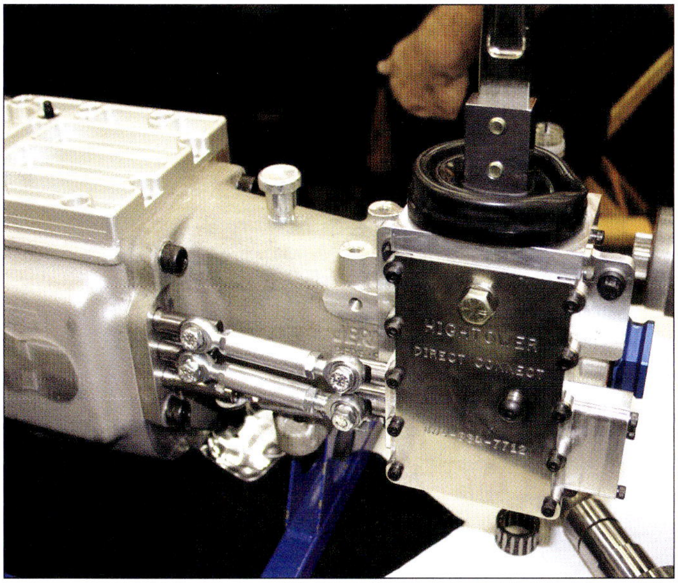

This Hightower Racing transmission can actuate shift rods that are completely vertical, which makes this shifter unique and keeps the rods closer to the transmission's extension housing.

The G-Force GT-50 clutchless front-wheel-drive drag race transmission uses "Remmed" gears and a special Teflon dry-film coating on the case to promote oil flow.

The Muncie first gear on the right has been shot peened, deburred, and Mikronited.

The 5speeds.com Muncie SPEC25 uses the same coating as the G-Force GT-50. It's a fully synchronized transmission for race classes that require a synchronized transmission. The gears are also processed in the same fashion.

CHAPTER 9

Clutch Basics

Although the focus of this book is manual transmissions, without proper clutch performance a transmission can be rendered useless. The premature failure of most synchronizers usually stems from some sort of clutch-related problem.

On any given week, I'll get a few calls about some sort of shifting issue. I'll ask the question, "Do you know how the clutch works?" The replies could fill a few chapters and appear humorous to some people, but the most common reply is "No."

McLeod Industries produces an extremely light multi-disc clutch and flywheel.

Think of the clutch as a simple on-and-off switch. Without it, the engine is connected to the transmission at all times. Power flow from the engine to the transmission is never interrupted, and therefore, transmission gears and components would be loaded at all times. In order for a manual-transmission synchronizer to work properly, power flow has to be interrupted before and during the shift so that the synchronizer can work freely. This break in the power flow occurs when you press down on your clutch pedal. You turn the switch off. Now the synchronizer can match gear speeds to the transmission output shaft with little effort. When this is accomplished, the shift is made, and power is reapplied by turning the switch on (lifting up on the clutch pedal).

Poor clutch release is the major cause of manual transmission problems. Your on-and-off switch has become shorted. Frequent causes of poor release are simply not enough pedal travel or effective throw to cause a proper release.

Clutch Components

Your clutch system is made up of six basic parts: the pressure plate, clutch disc, release bearing, flywheel, pilot bearing, and bell housing. The pilot bearing or pilot bushing is installed in your engine's crankshaft. Its job is to keep the transmission input shaft concentric with the crankshaft centerline while allowing it to spin inside of it. The bell housing locates the transmission to the engine. It also aids in keeping the transmission aligned with the engine. If you were to assemble a transmission to an engine with just a pilot bushing and bell housing, you would be able to spin the transmission freely while it maintains its concentricity with the crankshaft. The

Here is a McLeod hydraulic assembly mounted to a custom T5. Note that the line fittings swivel to allow you to fish the lines through the bell housing. Adjustment is made by changing pistons that come in several lengths.

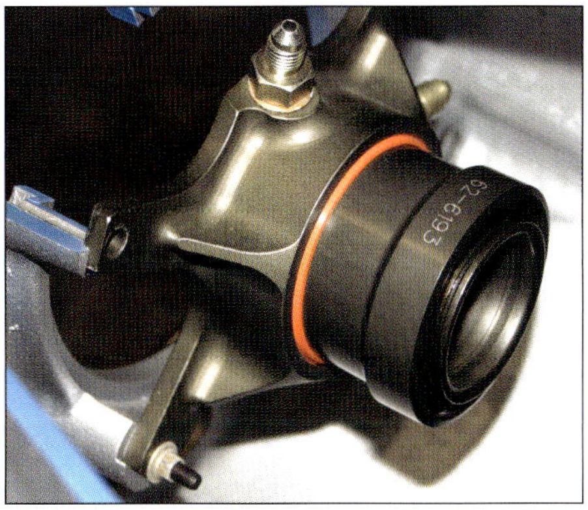

Here is a cutaway of a Tilton hydraulic release bearing that mounts inside of the bellhousing. Spacing it out or machining the housing provides adjustment.

bell housing is located to the block with alignment dowels. "Dialing-in" a bell housing is the term for centering a bell housing with the crankshaft. Offset dowels are often used to move the housing into spec. Most performance applications require run out to be less than 0.008 inch while maintaining a parallel surface of 0.005 inch. If the housing is out of alignment, the transmission input shaft will side load and bind on the crankshaft pilot bearing, causing pilot bushing and front transmission bearing failure as well as clutch chatter and release issues.

The flywheel stores rotational energy. Heavy wheels may cause an engine to accelerate more slowly but allow smoother shifts because rpm will not fluctuate much. A lighter wheel may allow the engine to rev faster but may cause jerky off-the-line clutch chatter because of RPM flutter.

Although most people don't associate release bearings or "throw out" bearings with the flywheel, they should. The flywheel is what the clutch pressure plate fastens to. If the height of the flywheel changes, the whole clutch assembly moves either closer or farther away from the release bearing. This is what typically happens when purchasing an aftermarket flywheel or having an older wheel resurfaced. Release bearings have pretty much been hydraulically activated since the mid 1980s. Resurfacing a flywheel with a hydraulic release bearing that has no adjustment causes most release issues. Mechanical linkages often compensate for these issues.

Clutch Types

At times it can be confusing as to what clutch you may need. Terms like "dual disc," "dual friction," "sprung hub," and "unsprung hub," don't make matters any easier.

I honestly don't understand most of the marketing hype surrounding clutches because a single-disc, organic-lined, diaphragm clutch has been used in many high-horsepower applications with a great life factor for many years. So why all the new technology?

The pressure plate, which is basically a clamp, ideally should have enough holding force for the intended application, yet be as easy to apply as possible. A certain amount of slippage should be required to avoid a jerky or harsh engagement. Harsh engagement can also lead to drivetrain breakage. The bigger the diameter the more the holding power, but the larger the diameter the more rotating mass. The new technology makes lighter, smaller clutches with great holding power with the least amount of pedal effort.

Think of applying pressure to your chest with your finger as opposed to your hand. The finger applies the same pressure but with more pounds per square inch. It is easier to rotate your little finger compared to the whole palm with the same pressure behind it, so you would need more

The full cutaway shows the bearing on the left against a triple disc 7-inch diameter racing clutch.

pressure to keep your finger rotating than your whole palm. This is the principle behind the design of a dual-friction disc.

Generally, a diaphragm clutch has less weight than a Long-style or Borg & Beck-style. Long and Borg & Beck clutches are also known as "3-finger" clutches because the diaphragm is both the lever and the spring. Borg & Beck and Long-Style clutches use three levers that push against a spring-loaded plate with coil springs. The Borg & Beck is not adjustable, while the Long style is. Pedal pressure will be easier with a diaphragm because a lever ratio of 10:1. A Borg & Beck has a typical ratio of 6:1 and the Long-style 4.8:1. Height is a consideration because obviously the taller the clutch the longer the transmission's input shaft will have to be.

Clutch discs come in a wide variety of sizes, hubs, and linings. Most street applications have clutch discs that absorb initial shock in two ways. The first is the drive hub, which is driven by springs, and the second is the surface the lining, which is bonded to a wavy spring-steel surface commonly called a marcel.

People often have a hard time telling if a pilot bushing, throw-out bearing, or clutch is bad. In most applications (other than a Long-style), you cannot adjust for a slipping clutch. No external adjustment is going to help. The best method for adjusting correct release is to measure clutch release with a feeler gauge. A typical good release is 0.040 inch. Some bell housings have a lower inspection cover. You can usually get a feeler gauge in between the flywheel and disc to get a measurement. This method is very accurate. Some people adjust a clutch by setting the amount of free play the pedal has before allowing the release bearing to make contact with the clutch. Using this method to adjust the clutch is actually the number-one cause of poor-release issues.

A quick down and dirty way to check for clutch release is to simply get the car good and warm so all components are heated and expanded. Press down on the clutch pedal and count to 10. Within 10 seconds the transmission should slow down on its own. Most transmissions (the ones discussed in this book, anyway) have an unsynchronized reverse. If you try to put the transmission into reverse after the 10-second wait and the gears grind or clash, the transmission is still turning and has not been decoupled from the engine.

The clutch drive plate will either be bolted or riveted to the lined surface. It is raised via stand offs to clear the sprung hub. The springs fit in the slots, which drive the plate. I'm removing the plate to expose the hub and springs.

Here you can see the sprung hub. It mounts to the transmission input shaft and drives the transmission via the springs.

Most new clutch designs are based on this typical diaphragm clutch like this one.

A Long-style clutch can be adjusted for more or less pressure.

Here are some Ferrari racing clutches. The one on the right is a solid clutch with no marcel and no springs. The hub is riveted solid to the lining.

Here is the sprung hub on a transmission input shaft. This simple visual can give you a better idea of how power flow is put through a clutch system.

The McLeod modular bellhousing allows for indicating the housing within itself because the mounting surface of the bellhousing is a separate component.

The Kevlar side of a dual-friction disc offers aggressive performance that's tame enough for street use.

The ClutchNet design locks the sprung hub to a solid drive plate. It uses fewer rivets and maintains better concentricity and strength. The springs act like little pistons absorbing shock loads.

The organic side of a dual-friction disc with poly encapsulated springs.

A Chevy bronze pilot bushing is on the left, and a factory roller-style pilot bearing is on the right. The needle-style bearing is less forgiving when a misalignment exists and can make a mess out of a crankshaft or input shaft if it fails.

Here is how a pilot bushing supports your input shaft. The correct support of this bushing and fit of a driveshaft yoke in the tail bushing are what keep things aligned.

Guide to Shifters

Today most manual-transmission cars are equipped with a floor shifter. In general, prior to World War II, most vehicles had a shifter mounted in the top cover of the transmission. The shifter directly moved the shift forks. Automotive engineers in Detroit wanted to get away from floor-mounted shifters and create shifters built into the steering column. In order to create a column-mounted shifter, the actual transmission shift mechanism now had to have a series of external shifter shafts, linkage arms, shift rods, and levers for this design to be feasible. These new components required a redesign of the actual transmission to incorporate the addition of all these extra components.

The column shifter design was a far cry from the positive feel of the direct-top cover shifter. To provide that positive shift feel, modifying column shift back to a floor shift became the thing to do when drag racing started to flourish in the late 1950s. Drag racing puts a tremendous strain on linkage components, and a common modification was usually a custom floor shifter. Companies, such as Fenton, Sparkomatic, P&G, Hurst and Mr. Gasket, were leaders in this field. The typical "H"-pattern shifter seems to have stood the test of time. The H-pattern design for the most part uses a horizontal gate. Thus, the shifter moves left or right to grab each gate. Bill Bieber is the person credited for the design of the Vertigate Shifter. The gates were set up in

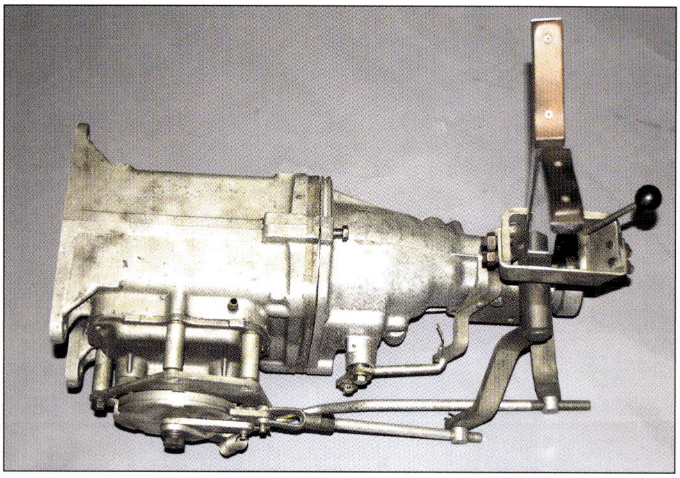

This is the first sequential-style shifter design ever made for a side-rail-shifted transmission. The driver pulls one lever for upshifts and another for downshifts.

Thin stampings and cast aluminum bodies made these early "hot-rod"-style shifters very weak.

a vertical alignment. This eliminated the need to move the shifter left or right. You just pulled the stick front to back to change gears. He also created a wild shifter, called "The Changer," which had two sticks. One was for upshifts and one for downshifts. You just pulled the sticks towards you to change gears. It used a huge rotating cam on the side cover. Most of these designs suffered in the durability department, as well. The very popular Fenton and Sparkomatic incorporated diecast aluminum bodies in some models, which quickly cracked under heavy use. The radically designed Changer required a huge portion of your floor pan to be cut out. Hurst and Mr. Gasket were the only two companies to survive the "shifter wars."

The Hurst Competition Plus shifter improved on all previous designs, and it's still one of the best aftermarket shifters.

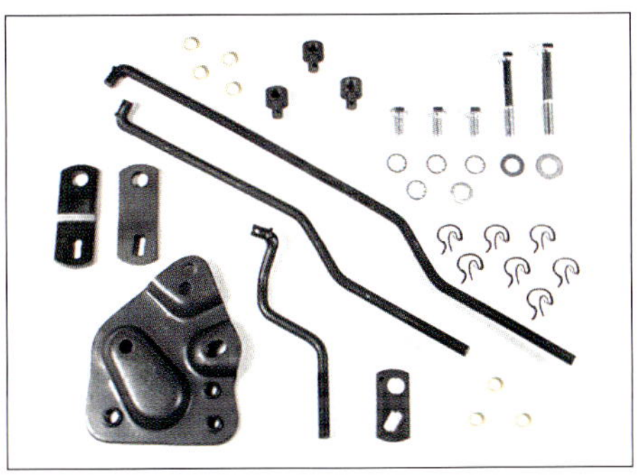

Hurst Shifters have a vast amount of installation kits for just about every side mounted 3- and 4-speed shifter.

Various patents were issued protecting most of these unique shifter designs, and therefore only a handful of aftermarket companies specialized in this field. Today, things are much different. The T5 transmission has a multitude of companies making replacement shifters, and almost every high-end transmission manufacturer has its own shifter.

The Need for Shifters

All conventional factory linkage-style shifters lack the use of a stopping mechanism. For normal applications a "stop" may not be needed. When the "stop" isn't present, the force of your shift is stopped internally by the detents of the transmission. This leads to premature failure of shift detents, shift forks, or shift lugs. The stopping device is usually a bolt that is adjusted to absorb the force of the shift thus extending the life of the internal components.

Improvement of shift "feel" is another important upgrade one should expect from a high-performance shifter. Usually they are designed to decrease travel lengthwise (short throw) and sometimes narrow the H-pattern. Others may incorporate a better left and right bias of the shifter. The factory GM Muncie 4-speed shifter had no spring mechanism in the shifter other than to lift the reverse lock out rod up. Since it could flop left or right, it was a clumsy to speed shift. You never knew the actual location of the shift gates. The Hurst Competition Plus design has stood the test of time. It has a perfectly designed bias that keeps the handle in the 3-4 gate, so it aids you when going from second into third during speed shifting. It has an additional spring force to overcome when engaging reverse, eliminating reverse lockout levers. GM actually started to install these shifters in the 4-4-2 and GTO.

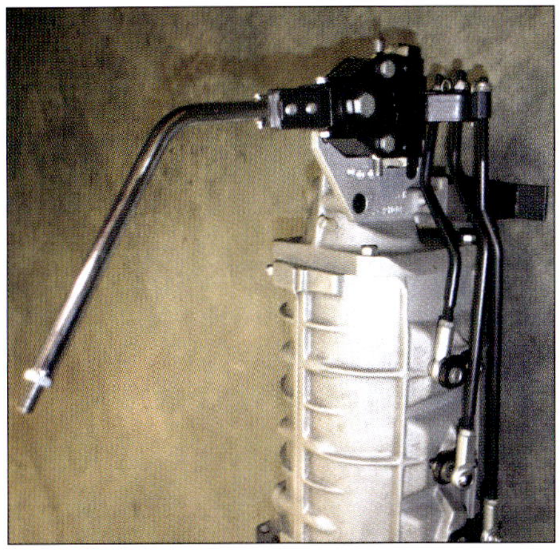

Here is a Long HN-1000 H-Pattern shifter designed for a Cobra kit car and bolted to a Richmond 5-speed.

The Long HN-1000 uses rails that run parallel with one another and in a straight line, as opposed to the Hurst shifter that has the gates move in an arc within the shifter body.

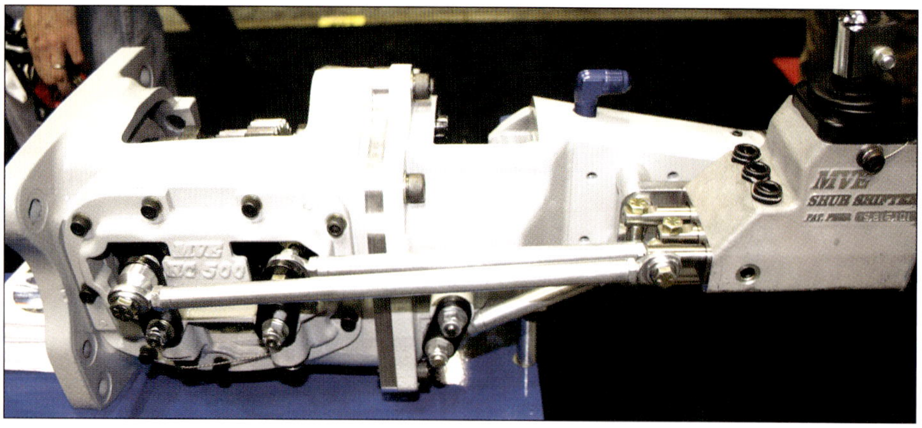

When it comes to road racing, Mid Valley Engineering's Shur Shifter is hard to beat. The whole mechanism is in a sealed housing. Basically, internal rails shift many transmissions, and the Shur Shifter can turn any external-linkage-designed transmission into a rail-shifted type with built-in stops and interlocks.

Sometimes in drag racing, the speed of your shift can make the difference between a win and a loss. These shifters were basically designed to create rapid shifts without the possibility of a "missed" shift.

Fixing the Hurst Competition Plus Shifter

There are more Hurst Competition Plus shifters floating around used and new than any other shifter around. By now, cleaning out the older units may be needed. These shifters, if properly maintained, will never wear out. Hurst still sells typical bushings and clips in what they call a Pit Pack.

Oddly, many people don't understand what shifter adjustment will and will not do. There are only two adjustments that can be made with any conventional Hurst Competition Plus shifter. The first is a neutral gate adjustment, and the second is stop adjustment. If your transmission is falling out of gear and you have a good neutral gate (a clean left-to-right motion of the shifter), you can't turn any rods to adjust away that problem. Stop bolts can get loose and work themselves in too far. If the shifter falls out of gear, remove the stops and see if this resolves the problem. To align these shifters a 1/4-inch rod is pushed through the shifter body and the three gates.

The dotted line shows the path your 1/4-inch diameter rod should take. The circle shows how the gates are aligned, but as you can see, they are not centralized within the shifter body. This is the only other way a shifter can cause a transmission to fall out of gear with a good neutral gate.

This is Long Shifters' high-tech vertical gated shifter. By pulling the lever up, your shifter is positioned with the top gate. Releasing the lever allows the springs to pull the shifter's arm vertically down each gate and eliminates the need for a conventional H-pattern.

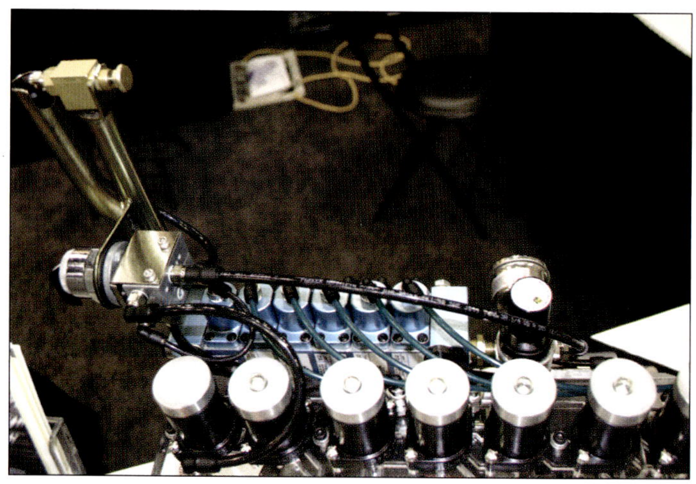

This is Jerico Transmission's air shifter. Solenoid valves route a charged high-pressure volume of air from one passage to another. A push-button actuates each shift, and a dial selects neutral or reverse.

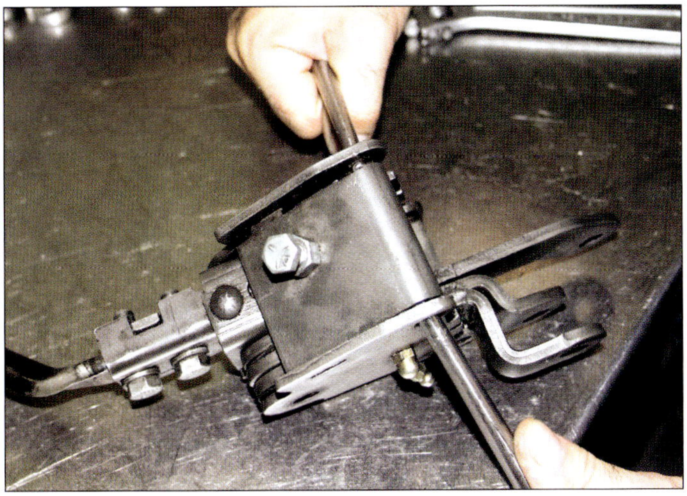

With all the bolts removed, use a pair of drift punches and pry the shift body apart.

Remove the back wall of the shifter body.

Remove the main pivot shaft by tapping it out or walking it out with a set of pliers.

Grasp the whole gate assembly and pull it out of the shifter body. Try keeping it together at this point.

Once out of the shifter body, carefully lay each piece down as pictured. You may want to clean each piece and return it to its proper place on the bench.

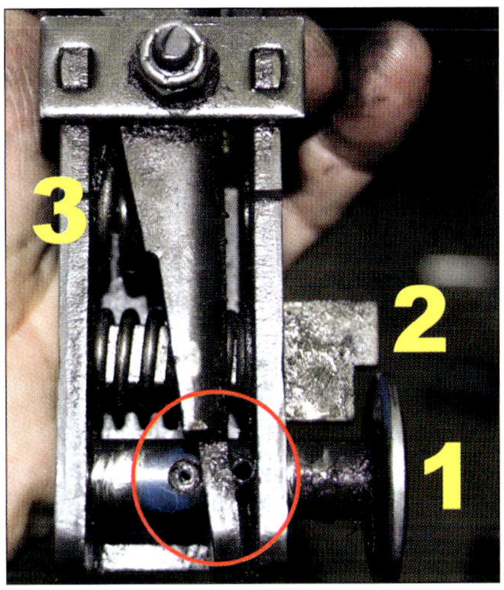

1) This is the plunger that basically contacts all three gates depending on how the shifter is biased left or right. 2) This internal stop is what gives you the extra effort to get into reverse. 3) This is the spring that handles the normal left-to-right bias. The circle shows the roll pins that push the plunger left or right. The pin on the left broke, allowing the stick to move toward first gear, but not the plunger.

Replace the factory roll pins with solid hardened pins available from any machine tool supply house. This is basically the only trick mod that needs to be done. Everything is cleaned. Reverse the above steps to assemble the shifter.

Notice how the McLeod shifter uses a spherical bearing. This eliminates the need to keep the pivot loaded, prevents pivot wear, and allows for total spring bias adjustment with Bellville washers.

The rod aligns the gates with each other and centralizes the gates within the shifter body.

T5/T56 Shifter Modifications

The factory T5 shifter is simply a lever that pivots on a ball socket.

Side-to-side and forward-and-backward motion both ride on this ball socket. The centerline distance of your pivot-ball socket to the level end centerline determines the length of your throw. Until 2004, every aftermarket T5 shifter used the same principle. They just housed them differently, or used their own variations of bias spring pressures to come up with their own feel.

McLeod Industries' Red Roberts designed a pivot system using a vertically mounted spherical bearing. The Slik Stix shifter's design creates a very efficient geometry that produces shorter throws and narrower gates within a smaller package.

The low-profile shape of the McLeod Shifter also allows for many offsets without creating floor pan interference.

Shown are a wide variety of factory shifters. I keep to mix and match parts for various conversion projects.

The T5 shifter is the most reproduced aftermarket shifter today. When it comes to the factory T5 shifter, there are no adjustments because it has only one internal shift rail, and neutral is controlled by an internal transmission track plate. It becomes a shifter with very few components. Hurst originally came up with a cast-alloy low-profile, short-throw shifter for Ford's SVO program. The shifter sparked interest from several companies to create billet shifters that didn't wear out as quickly. These sport billet-machined bodies with stop bolts. The stops keep the internal shift lugs from deforming under heavy load. Because factory T5 shifters are a non-serviceable item, I will show you how to mod the factory shifter for any specialized conversion you may have in mind.

Modifying the T5/T56 shifters in this manner makes for a cleaner-looking piece as opposed to some stick that is bolted on and spaced out in every direction possible. It also makes for a better-shifting assembly because the geometry is much more direct.

A stock T5 GM-style shifter is shown on the left, and a billet Steeda Tri-Ax shifter appears on the right.

Start to dismantle the T5 shifter by cutting off the rear folded section. Unfolding it and refolding will only weaken it, so just cut it off.

Tap the cover rearward with a hammer, and the whole shifter will come apart. Lift the stub assembly out of the socket base, and remember the order of the flat springs. Slide off the nylon ball.

I can now start welding since all plastic parts and springs are removed. I'm welding a special fitting for my JT5 Jaguar shifter.

The fitting creates the proper offset any angle for my conversion. You can do the same for any other conversion. The concept is the same.

A normal-throw shifter stub is left, while a short-throw stub is on the right. See how they are the same exact length, but the position of the pivot shaft is higher on the short-throw version. The base is also higher on the short-throw shifter. You cannot mix bases and stubs, meaning a short-throw base stays with a short-throw stub.

Assemble the shifter with all internal components greased. Clamp it together in a vise along with some Vise-Grips clamping the top half together with the base. It acts as a heat sink while the two halves are fuse-welded together. If you make a really long stick, add a couple of extra flat bias springs to get a better bias feel. Shifter internal parts are not available, so that is why I keep lots of shifters around.

The fitting I weld on allows me to screw on our own custom stick, shift ball and locking ferule.

Common General Motors Speedometer Gears

Part #	Descriptions
#345215	GM Speedometer fitting with seal
#3708148	GM Hold Down

Mainshaft Drive Gear (Still Available)

Part # **Tooth Count**

These fit a 27-Spline output shaft

3708144	8T	1.84-inch OD
3708145	8T	1.76-inch OD

This fits a 32-Spline output shaft

3978758	8T	1.84-inch OD

Mainshaft Driven Gear (Still Available)

Part # **Tooth Count**

These mate with the 1.84-inch drive gear.

3987917	17T – Purple
3987918	18T – Brown
3987019	19T – Neutral/White
3987920	20T – Blue
3987921	21T – Red
3987922	22T – Gray/Black

These mate with the 1.76-inch drive gear and are larger in diameter than the above gears.

3860345	22T – Green
3860346	23T – Black
3860347	24T – Yellow
3860348	25T – Orange

Autogear has produced 7-tooth versions of the large-diameter versions of both spline speedo-drive gears.

Part Numbers	Tooth Count
18-110-001	7T – 27-Spline Drive
18-110-002	7T – 32-Spline Drive

Speedometer Gear Ratio Formula

This formula can be used to determine the proper speedometer-driven gear.

Speedometer Gear Ratio (SGR) = (63360 x Axle Ratio) / (3141.6 x Tire Diameter)

(SGR x Speedo Drive Tooth Count) = Number of driven gear teeth.

Torque Specifications and Oil Capacities

Oil Types

The Internet has created many online enthusiast forums for cars and transmissions. If you have visited one of these forums, at least a few times a week someone is asking about what oil type or brand they should use in their transmission. While there are many self-proclaimed experts residing in these forums, many do not provide reliable or accurate information. There has been a marketing hype by synthetic oil manufacturers claiming that the current 80W-90 weight API - GL5 oil will corrode bronze synchro rings. This may be true in theory, but after rebuilding thousands of transmissions, I have never seen this happen. Synthetics have a great deal of merit in cold climates. Transmissions, such as the Muncie, that don't have a front seal, usually end up leaking oil because synthetic oil flows better than conventional gear lube. In my experience, any 80W-90 or 75W-90 seems to work well in the Muncie, T10, Super T10, and Toploader. The A833 Chrysler without bushed gears uses gear lube as well. The T5 and A833 with bushed gears all use Dextron III-style lubricant.

Oil Capacity

People seem to be obsessed with oil capacity. None of these manual transmissions discussed in this book has a dipstick. Filling the transmission up to the level of the fill plug determines the correct lubricant capacity. So the only reason you would need to know the capacity is to know how much oil you will need to purchase. Transmissions, such as the T5 and A833, have different-length extension housings, and therefore capacity can vary within the same brand. Muncie and T10-style transmissions usually take 1-1/4 to 1-1/2 quarts, while T5 and A833 transmissions accept close to three quarts. So if you pick up three quarts of oil, you'll probably be covered and have a little extra just in case you have a spill.

Muncie 4-Speed – Torque Specs

Side Cover	18 ft-lbs
Front Retainer	25 ft-lbs
Extension Housing	30 ft-lbs
Speedo Hold Down	10 ft-lbs
Fill/Drain Plugs	25 ft-lbs

T10 And Super T10 – Torque Specs

Side Cover	18 ft-lbs
Front Retainer	18 ft-lbs
Extension Housing	30 ft-lbs
Speedo Hold Down	10 ft-lbs
Fill/Drain Plugs	25 ft-lbs
Mid Plate Bolt	30 ft-lbs

T5 5-Speed – Torque Specs (NWC and WC)

Front Retainer	18 ft-lbs
Countergear Lock Plate	18 ft-lbs
Top Cover	10 ft-lbs
Shifter Base	18 ft-lbs
Extension Housing	25 ft-lbs
Fill/Drain Plug	25 ft-lbs
Hex Pivot	35 ft-lbs
Speedo Hold Down	10 ft-lbs

Toploader 4-Speed – Torque Specs

Top Cover	18 ft-lbs
Front Retainer	18 ft-lbs
Extension Housing	30 ft-lbs
Speedo Hold Down	10 ft-lbs
3-4 Detent Bolt	20 ft-lbs
Fill/Drain Plug	25 ft-lbs

Chrysler A833 4-Speed – Torque Specs

Side Cover	18 ft-lbs
Front Retainer	18 ft-lbs
Extension Housing	30 ft-lbs
Speedo Hold Down	10 ft-lbs
Fill/Drain Plug	20 ft-lbs

Ratios at a Glance

A quick guide to of the popular and not so popular gear ratios.

Ford Toploader

1ST	2ND	3RD	4TH
2.32	1.69	1.29	1.00 Close Ratio
2.78	1.93	1.35	1.00 Wide Ratio

Muncie 4-Speed

1ST	2ND	3RD	4TH	
2.20	1.64	1.28	1.00	M21/M22 Close Ratio
2.56	1.91	1.48	1.00	M20 Early Wide Ratio
2.52	1.88	1.46	1.00	M20 Late Wide Ratio
2.52	1.73	1.35	1.00	M21W Autogear Wide Ratio
2.56	1.75	1.36	1.00	M22W Autogear Wide Ratio
2.20	1.51	1.17	1.00	M22X Autogear Extra Close Ratio
2.20	1.51	1.00	0.86	M21Y Autogear Overdrive Ratio

T10 4-speed (1957 – 1965)

Series	1ST	2ND	3RD	4TH	
T10	2.20	1.66	1.31	1.00	
T10A	2.54	1.92	1.51	1.00	
T10B	2.36	1.78	1.41	1.00	Ford
T10C	2.20	1.64	1.31	1.00	
T10D	2.54	1.89	1.51	1.00	
T10E	Same as T10D but with SAE 9310 gears				
T10H	2.74	2.04	1.50	1.00	Ford
T10J	2.36	1.76	1.41	1.00	Ford
T10K	Same as T10C but with SAE 9310 gears				
T10L	2.36	1.76	1.41	1.00	Ford
T10M	2.36	1.62	1.20	1.00	Ford Shelby
T10N	2.54	1.74	1.30	1.00	Aftermarket

Super T10 4-speed (1st Design)

T10U	3.44	2.28	1.46	1.00	GM Truck
T10W	2.64	1.75	1.33	1.00	
T10S	2.43	1.61	1.23	1.00	
T10P	2.43	1.76	1.47	1.00	AMC
T10Q	2.64	2.10	1.60	1.00	FORD
T10T	2.64	2.10	1.46	1.00	AMC
T10V	2.23	1.77	1.35	1.00	AMC

Super T10 4-speed (2ND Design Only)

T10W	2.64	1.75	1.33	1.00	
T10X	2.64	1.61	1.23	1.00	
T10Y	2.88	1.75	1.33	1.00	
T10Z	3.42	2.28	1.46	1.00	
T10AA	3.29	1.72	1.00	.80	Rare Overdrive Set
T10BB	4.17	2.28	1.46	1.00	
T10CC	2.88	1.91	1.33	1.00	GM Version

Saginaw 4-speed

1ST	2ND	3RD	4TH
3.50	2.48	1.65	1.00
3.11	2.20	1.47	1.00
2.85	2.01	1.47	1.00
2.54	1.80	1.44	1.00